ASIMOV AT HIS BEST

- How the universe got here—and how it will end
- The fifteen elements to make an ocean—anywhere
- A future where all work is done at home
- The possibility of an infinite number of universes, separated by something more subtle than place and time
- Computers to design human evolution
- How lunar colonists can "hook" a comet as a source for water

"Not merely an author of science fiction, Isaac Asimov is one of the widest-ranging intellects at large in the world today. . . . Asimov is interesting because he is interested. It is refreshing to find a writer who can explore the issues with such clarity and precision."

—COLUMBUS DISPATCH

"Asimov is the dean of popular science writers."

—LOS ANGELES TIMES

Books by Isaac Asimov

Asimov on Numbers
The Beginning and the End
The Collapsing Universe

Published by POCKET BOOKS

Isaac Asimov

THE BEGINNING AND THE END

PUBLISHED BY POCKET BOOKS NEW YORK

POCKET BOOKS, a division of Simon & Schuster, Inc.
1230 Avenue of the Americas, New York, N.Y. 10020

Published by arrangement with Doubleday & Company, Inc.
Library of Congress Catalog Card Number: 76-52218

ISBN: 0-671-47644-0

First Pocket Books printing December, 1978

10 9 8 7 6 5 4 3

POCKET and colophon are registered trademarks
of Simon & Schuster, Inc.

Printed in the U.S.A.

ACKNOWLEDGMENTS

"The Real Cyrano" originally appeared as "Cyrano de Bergerac" in the
February 2, 1974, issue of *TV Guide®* Magazine, copyright © 1974
by Triangle Publications, Inc., Radnor, Pa. Reprinted with permission.

"Of What Use?" originally appeared as the introduction to *The Greatest
Adventure: Basic Research That Shapes Our Lives,* edited by E. H.
Kone and H. J. Jordan, copyright © 1973 by Rockefeller University
Press.

"The Democracy of Learning" is reprinted from Volume 1, Number 4,
of *Know,* copyright © 1974 by Encyclopaedia Britannica.

"The Monsters We Have Lived With" originally appeared as "They
Don't Make Monsters Like They Used To" in the November 23, 1974,
issue of *TV Guide®* Magazine, copyright © 1974 by Triangle Publications, Inc., Radnor, Pa. Reprinted with permission.

"Fossil Fuels" originally appeared as "The Fascinating Story of Fossil
Fuels" in the August–September 1973 issue of *National Wildlife,* copyright © 1973 by the National Wildlife Federation.

"A Drop of Water" is reprinted from the Spring 1976 issue of HTH
Poolife, copyright © 1976 by Olin Corporation.

"Smart, But Not Smart Enough?" originally appeared as "Chimpanzees"
in the May 4, 1974, issue of *TV Guide®* Magazine, copyright © 1974
by Triangle Publications, Inc., Radnor, Pa. Reprinted with permission.

"Recipe for an Ocean" originally appeared as "Recipe for a Planetary
Ocean" in the August–September 1976 issue of *Natural History,* copyright © 1976 by The American Museum of Natural History.

"Technology and Energy," under the title "Technology: Whale Oil,
Arab Oil and No Oil" was published as Number 5 in a series of
dialogues on technology by Gould Inc.

To Merrill Panitt
    Roger J. Youman
    William Marsano
    and all the other nice people of *TV Guide*

CONTENTS

INTRODUCTION

In the course of my writing career, so far, I have written over eight hundred nonfiction essays on dozens of different subjects. This may seem an unconscionable outpouring of words, and the world has a right to know on what grounds I feel justified in inflicting all of this on it. Well—

1. It gives me an innocent pleasure to do so. After all, I have a lot to say and trying to keep it all pent up within me may easily damage my internal organs.

2. I get paid for it, and I do have to earn a living.

3. These essays are read by others only as a voluntary act, so that I force harm on no one against his or her will.

It may be that you find these reasons satisfactory and are prepared to forgive me* but then I compound the mischief by periodically grabbing up an armful of manuscripts and organizing them into a collection. I have done this over and over—twenty-two times, actually—but I have reasons for this, too.

1. The original appearance of my articles in magazines or in other periodicals is evanescent. The article is there, and then in a day, or a week, or a month, it is gone. A book affords my articles greater permanence.

2. Each of my articles appears originally in the various periodicals embedded in extraneous matter. In a book you get only me, and nothing more.

3. Few people are likely to see all the various periodi-

---

* Incidentally, if you have never read anything by me before, and if this introduction is actually the first sample of my writing that you have seen, it is only fair to warn you that this personal way of communicating with you is characteristic of me. If you find it nauseating, please don't read any further. I would rather lose the royalty involved in the sale of this book to you than have you sick on my account.

cals in which my articles have appeared, so that a collection is bound to include numerous items a particular reader has missed, even if he happens to be a fan of mine.

These reasons assume, of course, that my articles deserve permanence, that there is an advantage to having them concentrated, and that missing one of my articles is a tragedy. Well, I *do* assume that, and I assume the reader does, too, or he wouldn't be reading this book.

To be sure, a collection such as this one involves difficulties. The chief of these arises out of the fact that I am willing to write on just about any subject, so that a collection of nearly two dozen essays is bound to be a pretty mixed bag. Some essays may be on subjects that don't interest a given reader at all.

My defense there is that where an essay subject is not interesting in itself to a particular reader, my writing style (which I do my best to make engaging) may cause him to *become* interested, and what's wrong with that?

Of course, I do my best to make sense out of the mixed bag by trying to arrange the essays in some sensible way that will lead you by gentle stages from one subject to another. In this book, for instance, I will go from the past through the present to the future.

A second difficulty is that I am frequently asked to write an article on a subject that I have dealt with before, with the result that there is always the danger of overlapping. Some of the articles in this book, for instance, overlap articles that have appeared in other collections of mine.

Since I usually take up a subject a second or third time from a new angle or in a new context, the overlapping is not as bad as it sounds. Besides, it may well be that you have never read the earlier article; or that, having read it, you may (unlikely as that may sound) have forgotten it.

Enough then, and let us start.

THE
PAST

1 THE REAL CYRANO

FOREWORD

Where do I get my information? Encyclopedias, textbooks, dictionaries, magazines, newspapers, and a lifetime of miscellaneous reading which my retentive memory has filed away somewhere in very neat order.

Generally, the material I marshal out in order to prepare an article is obtained in so diffuse a manner that it is almost impossible for me to list my sources in any reasonable way.

Not always, however. Sometimes, my source is very easily identifiable and then I am tempted to give credit. My reference for the following article, for instance, is the chapter on Cyrano in the book Explorers of the Infinite *(World, 1963), which was written by my good friend Sam Moskowitz.*

Lest you think I helped myself improperly, let me assure you of two things—first, the words and arrangement are my own, and, second, I called Sam and got his permission.

There must be millions of people who know that Cyrano de Bergerac is a fictional character in a play by Edmond Rostand, a character who was pictured as having a huge nose and who was forever killing people who mentioned the fact; one who was a poet, playwright, wit, romantic, and so foolhardy as to fight a duel with a hundred opponents at once and to beat them, too. What an extravagant image of an impossible hero!

But Cyrano really lived! He was a real person, who really had a huge nose (not as huge as shown in the play) and who really fought duels (perhaps not as extravagantly as in the play), including one with a hundred adversaries at once in which he was really the victor; and who was really a poet, a playwright, a wit, and a romantic (perhaps not to quite the extent shown in the play).

And yet the most important thing about Cyrano de Bergerac, the *real* Cyrano, is barely referred to in the play. It seems that in addition to all else, Cyrano was one of the earliest science fiction writers; the best, perhaps, up to the time of Jules Verne.

Savinien Cyrano de Bergerac (Cyrano is his *last* name) was born near Paris on March 6, 1619. In 1638, while he was still in his teens but had already achieved a formidable reputation as a swordsman, he joined the army. He remained in active service only a few years, however, for he was twice seriously wounded. In 1643 (still only twenty-four years old) he turned to writing as a profession.

He wrote several tragedies for the French stage, which were not as successful as they deserved to be, and from which the great French dramatist Molière condescended to borrow a few scenes. Cyrano's masterpiece, however, or at least that portion of his writings for which he is known today, is *A Voyage to the Moon,* of which an edition was smuggled into print without Cyrano's permission in 1650.

Cyrano died in Paris on July 28, 1655 (at the age of thirty-six), as a result of injuries from a beam which fell on his head—the accident having perhaps been arranged by men he had offended. After his death, the first authorized edition of *A Voyage to the Moon* was published in 1657, and in 1662 its sequel, *A Voyage to the Sun*, was also published. Both originally appeared in censored and mutilated form, but the original manuscript of the former, at least, survived, so that we now have *A Voyage to the Moon* in the form in which Cyrano wrote it.

The most interesting thing about the book and its sequel is Cyrano's description of methods for reaching the Moon. Until then, fictional trips to the Moon had been made by way of waterspouts, by vehicles hitched to flocks of birds, and by the use of demons or angels for transport. Cyrano tried to describe methods never before used, and succeeded in listing no fewer than seven!

1. You could cover yourself with bone marrow. Since it was believed at the time that the Moon sucked up bone marrow, you would rise with the marrow to the Moon. (This is pure legend, of course, and worthless.)

2. You could strap vials of dew to your body, and rise when the dew rose. (Dew *does* rise, unlike bone marrow, but only because it vaporizes. Even as vapor, it rises at most only a few miles up into the air—and it doesn't vaporize at all if it is in a closed vial.)

3. You could stand on an iron plate, holding a magnet. Throw the magnet into the air; the plate will follow. Catch the magnet and throw it up again. In this way, the plate will follow the magnet as high as your muscles will hold out, and all the way to the Moon, perhaps. (This one sounds good, but it won't work. Imagine yourself in mid-air. If you throw the magnet upward, you drive the plate downward by just enough to cancel the following upward motion, and the whole system is pulled down to the Earth by gravity.)

4. You could fill an airtight container with air, heat it, and force it out of an opening. (This is a vision of the principle of the jet plane, and of course it would work only as long as air surrounded the container. In Cyrano's time, however, it was just beginning to be understood that air was limited to the immediate neighborhood of the Earth and that between Earth and Moon there stretched many thousands of miles of vacuum. Until Cyrano's time it had been taken for granted that air filled the entire space between Earth and Moon.)

5. You could fill a globe with smoke, and since smoke ascends, the globe, carrying a passenger, could rise to the Moon. (This is a vision of the balloon but again this would work only if air filled the entire space between Earth and Moon.)

6. You could devise a machine like a large steel grasshopper and keep it leaping through the air by means of

explosions of gunpowder. (This is a vision of the airplane, but what Cyrano needed was a propeller in place of legs, and flat wings.)

None of these six methods will work, you see, either because they are based on myth, or are mechanically impossible, or are helpless in a vacuum. But then there is a seventh method described.

7. You can attach rockets to your vessels.

As it happens, this was the way in which the hero of *A Voyage to the Moon* begins his trip through space. Some French soldiers in Canada strap rockets to his vessel as a joke, and touch them off. High up into the air goes the spaceship, and when the rocket impulse dies, bone marrow which the hero has rubbed all over himself takes him the rest of the way.

As it turns out, the rocket principle *can* carry a man to the Moon; it is the technique by which men were, in actual fact, carried to the Moon in 1969.

This is a remarkable bit of intuition on Cyrano's part, since the rocket principle depends on the Third Law of Motion, which was first announced by Isaac Newton in 1687, long after Cyrano's death.

After the appearance of Newton's great book in which he described the three laws of motion and worked out the law of universal gravitation, it was easy to see that the rocket principle would work, and why, too.

But Cyrano was there before Newton by nearly forty years. He was there before *anybody*

When Neil Armstrong's foot touched the Moon, marking that giant leap for mankind, somewhere in the Valhalla reserved for science fiction writers, the spirit of Cyrano must have smiled. It had taken three centuries, but the most remarkable single insight in the history of science fiction had come true.

2 OF WHAT USE?

FOREWORD

The following article appeared originally in the form of an introduction to a book on modern research. The title of the book, The Greatest Adventure, *was taken from the last line of this introduction.*

I have never counted up how many introductions I have done for books other than my own. I don't know what the record is for any one author, but in my more melancholy moments, when I have once again been talked into doing one, I think I must have broken that record long ago.

The only way to make the task bearable is to refuse to allow the introduction to be a mere encomium on the book. Instead, I try to make them independent essays that make points other than those made in the book.

It is the fate of the scientist to face the constant demand that he show his learning to have some "practical use." Yet it may not be of any interest to him to have such a "practical use" exist; he may feel that the delight of learning, of understanding, of probing the Universe is its own reward entirely. In that case, he might even allow himself the indulgence of contempt for anyone who asks more.

There is a famous story of a student who asked the Greek philosopher Plato, about 370 B.C., of what use were the elaborate and abstract theorems he was being taught. Plato at once ordered a slave to give the student a small coin so that he might not think he had gained knowledge for nothing, then had him dismissed from the school.

The student need not have asked, and Plato need not have scorned. Who would today doubt that mathematics has its uses? Mathematical theorems which seem unbearably refined and remote from anything a sensible man can have any interest in turn out to be absolutely necessary to such highly essential parts of our modern life as, for instance, the telephone network that knits the world together.

This story of Plato, famous for two thousand years, has not made the matter plainer to most people. Unless the application of a new discovery is clear and present, most are dubious of its value.

There is a story of the English scientist Michael Faraday that illustrates this. He was in his time an enormously popular lecturer as well as a physicist and chemist of the first rank. In one of his lectures in the 1840s, he illustrated the peculiar behavior of a magnet and a spiral coil of wire which was connected to a galvanometer that would record the presence of an electric current.

There was no current in the wire to begin with, but when the magnet was thrust into the hollow center of the spiral coil, the needle of the galvanometer moved to one side of the scale, showing that a current was flowing. When the magnet was withdrawn from the coil, the needle flipped in the other direction, showing that the current was now flowing the other way. When the magnet was held motionless in any position within the coil, there was no current at all, and the needle was motionless.

At the conclusion of the lecture, one member of the audience approached Faraday and said, "Mr. Faraday, the behavior of the magnet and the coil of wire was interesting, but of what possible use can it be?"

And Faraday answered politely, "Sir, of what use is a newborn baby?"

It was precisely the phenomenon whose use was questioned so peremptorily by one of the audience which Faraday made use of to develop the electric generator, which, for the first time, made it possible to produce electricity cheaply and in quantity. That, in turn, made it possible to build the electrified technology that surrounds us today and without which life, in the modern sense, is inconceivable. Faraday's demonstration was a newborn baby that grew into a giant.

Even the shrewdest of men cannot always judge what is useful and what is not. There never was a man so ingeniously practical in judging the useful as Thomas Alva Edison, surely the greatest inventor who ever lived, and we can take him as our example.

In 1868, he patented his first invention. It was a device to record votes mechanically. By using it, congressmen could press a button and all their votes would be instantly recorded and totaled. There was no question but that the invention worked; it remained only to sell it. A congressman whom Edison consulted, however, told him, with mingled amusement and horror, that there wasn't a chance of the invention's being accepted, however unfailingly it might work.

A slow vote, it seemed, was sometimes a political necessity. Some congressmen might have their opinions changed in the course of a slow vote where a quick vote might, in a moment of emotion, commit Congress to something undesirable.

Edison, chagrined, learned his lesson. After that, he decided never to invent anything unless he was sure it would be needed and wanted and not merely because it worked.

He stuck to that. Before he died, he had obtained nearly 1300 patents—300 of them over a four-year stretch, or one every five days, on the average. Always, he was guided by his notion of the useful and the practical.

On October 21, 1879, he produced the first practical electric light, perhaps the most astonishing of all his inventions. (We need only sit by candlelight for a while during a power breakdown to discover how much we accept, and take for granted, the electric light.)

In succeeding years, Edison labored to improve the electric light and, mainly, to find ways of making the glowing filament last longer before breaking. As was usual with him, he tried everything he could think of. One of his hit-and-miss efforts was to seal a metal wire into the evacuated electric light bulb, near the filament but not

21

touching it. The two were separated by a small gap of vacuum.

Edison then turned on the electric current to see if the presence of a metal wire would somehow preserve the life of the glowing filament. It didn't and Edison abandoned the approach. However, he could not help noticing that an electric current seemed to flow from the filament to the wire across that vacuum gap.

Nothing in Edison's vast practical knowledge of electricity explained that, and all Edison could do was to observe it, write it up in his notebooks, and, in 1884 (being Edison), patent it. The phenomenon was called the "Edison effect" and it was Edison's only discovery in pure science.

Edison could see no use for it. He therefore pursued the matter no further and let it go, while he continued the chase for what he considered the useful and practical.

In the 1880s and 1890s, however, scientists who pursued "useless" knowledge for its own sake discovered that subatomic particles (eventually called "electrons") existed, and that the electric current was accompanied by a flow of electrons. The Edison effect was the result of the ability of electrons, under certain conditions, to travel, unimpeded, through a vacuum.

In 1904, the English electrical engineer John Ambrose Fleming (who had worked in Edison's London office in the 1880s in connection with the developing electric light industry) made use of the Edison effect and of the new understanding which the electron theory had brought, and devised an evacuated glass bulb with a filament and wire which would let current pass through in one direction and not in the other. The result was a "current rectifier."

In 1906, the American inventor Lee De Forest made a further elaboration of Fleming's device, introducing a metal plate which enabled it to amplify electric current as well as rectify it. The result is called a "radio tube" by Americans.

It is called that because it was only such a device that could handle an electric current with sufficient rapidity and delicacy to make the radio a practical device for receiving and transmitting sound carried by the fluctuating amplitude of radio waves.

In fact, the radio tube made all of our modern electronic devices possible, including television.

The Edison effect, then, which the practical Edison

shrugged off as interesting but useless, turned out to have more astonishing results than any of his practical devices. In a power breakdown, candles and kerosene lamps can substitute (however poorly) for the electric light, but what substitute is there for a television screen? We can live without it (if we consider it only an entertainment device, which does it wrong), but not many people seem to want to.

In fact, the problem isn't a matter of showing that pure science can be useful. It is much more difficult a problem to find some branch of science that *isn't* useful. Between 1900 and 1930, for instance, theoretical physics underwent a revolution. The theory of relativity and the development of quantum mechanics led to a new and more subtle understanding of the basic laws of the universe and of the behavior of the inner components of the atom.

None of it seemed to have the slightest use to mankind, and the scientists involved, a brilliant group of young men, had apparently found an ivory tower for themselves which nothing could disturb. Those who survived into later decades looked back on that happy time of abstraction and impracticality as a Garden of Eden out of which they had been evicted.

For out of that abstract work, there unexpectedly came the nuclear bomb, and a world that lives in terror, now, of a possible war that could destroy mankind in a day.

But it did not bring only terror. Out of that work, there came radioisotopes which have made it possible to probe the workings of living tissue with a delicacy otherwise quite impossible, and whose findings have revolutionized medicine in a thousand ways. There are also nuclear power stations which, at present and in the future, offer mankind the brightest hope of ample energy during all his future existence on Earth.

There is nothing, it turns out, that is more practical, more downright important to the average man, whether for good or for evil, than the ivory tower researches of the young men of the early twentieth century who could see no use in what they were doing and were glad of it, for they wanted only to revel in knowledge for its own sake.

The point is, we cannot foresee the consequences in detail. Plato, in demonstrating the theorems of geometry, did not envisage a computerized society. Faraday knew that his magnet-born electric current was a newborn baby, but he surely did not foresee our electrified technology. Edison certainly didn't foresee a television set when he puzzled over the electric current that leaped the vacuum, and Einstein, when he worked out the equation $E = mc^2$ from purely theoretical considerations in 1905, did not sense the mushroom cloud as he did so.

We can only make the general rule that through all of history, an increased understanding of the Universe, however out-of-the-way a particular bit of new understanding may seem, however ethereal, however abstract, however useless, has always ended in some practical application (even if sometimes only indirectly).

It cannot be predicted what the application will be in advance, but we can be sure that it will have both its beneficial and its uncomfortable aspects. (The discovery of the germ theory of disease by Louis Pasteur in the 1860s was the greatest single advance ever made in medicine and led to the saving of countless millions of lives. Who can quarrel with that? Yet it has also led, in great measure, to the dangerous population explosion of today.)

It remains for the wisdom of mankind to make the decisions by which advancing knowledge will be used well and not ill, but all the wisdom of mankind will never improve the material lot of man unless advancing knowledge presents it with the matters over which it can make those decisions. And when, despite the most careful decisions, there come dangerous side effects of the new knowledge —it is only still further advances in knowledge that will offer hope for correction.

And now we stand in the closing decades of the twentieth century, with science advancing as never before in all sorts of odd, and sometimes apparently useless, ways. We've discovered quasars and pulsars in the distant heavens. Of what use are they to the average man? Astronauts have brought back Moon rocks at great expense. So what? Scientists discover new compounds, develop new theories, work out new mathematical complexities. What for? What's in it for you?

No one knows what's in it for you right now, any more than Plato knew in his time, or Faraday knew, or Edison knew, or Einstein knew.

But *you* will know if you live long enough; and if not, your children or grandchildren will know. And they will smile at those who say, "But what is the use of sending

24

rockets into space?" just as we now smile at the person who asked Faraday the use of his demonstration.

In fact, unless we continue with science and gather knowledge, whether it is seemingly useful on the spot or not, we will be buried under our problems and find no way out.

It is up to you, then, and up to everyone, to support science and, where possible, to keep abreast of it, for today's science is tomorrow's solution—and tomorrow's problems, too—and most of all, mankind's greatest adventure, now and forever.

AFTERWORD

"Of What Use?" was the occasion for a pleasant surprise. I received a copy of the November 1974 Reader's Digest in the mail. Since I have no subscription to the magazine, I wondered why it had been sent to me. Filled with a wild surmise, I leafed through it carefully and, almost at once, came across my own name. "Of What Use?" had been reprinted there and the publishers had not bothered to tell me of the sale. (Fear not, Gentle Reader, I received appropriate payment.)

3 THE DEMOCRACY OF LEARNING

FOREWORD

In 1974, when the new edition of the Encyclopaedia Britannica *came out (with myself included as an item in the "Micropaedia" volumes), the* Britannica *put out a magazine called* Know, *which was intended to be about the* Encyclopaedia *in its various aspects.*

Know was running articles by people noted for some facet or other of a busy intellectual life, and each article was to deal with some aspect of the usefulness of the Encyclopaedia. *Three articles of this type had been written, one for each of the first three issues. For the fourth issue, the editor asked for an article from me.*

With my own self an item in the Encyclopaedia *was I going to refuse any reasonable request they might make of me? Of course not. I wrote it.*

Euclid, it is said, was once engaged in demonstrating some of the propositions of geometry to King Ptolemy of Egypt. At last Ptolemy sighed and said, "Is there no way,

Eukleides, in which these demonstrations of yours can be made less finicky and complicated?"

"Ptolemaios," said Euclid severely, "there is no royal road to learning."

But Euclid, in upholding the democracy of learning—in saying that there was but one route that all, from beggar to ruler, must follow—was wrong. There *is* a royal road to learning, as the following story will indicate.

In the old days of France's kings, a mathematician was hired to instill the rudiments of geometry into the young head of a scion of one of the land's most ancient ducal families.

Politely, the young man listened to the mathematician's steady progress from step to step until the proposition was triumphantly demonstrated. Then he shook his head, took a pinch of snuff, and said, "Alas, monsieur, I do not see that."

The mathematician's lips pressed together and then he went over each step very carefully, explaining the logic meticulously and again ending with a splendid proof.

And again, the young man shook his head and, smiling agreeably, said, "But, monsieur, I still do not see that."

Whereupon the mathematician, suppressing a groan, said, "Monsieur le duc, I give you my *word* that what I have said is so."

"Ah, monsieur," said the young man with a bow, "why did you not say so at once? I would never permit myself the liberty of doubting you."

And *that* is the royal road to learning—the acceptance of authority. There is nothing so easy as to accept, unquestioningly, everything you are told. Learn to parrot it, and you may even manage to sound quite learned. But where's the fun in it?

During my Brooklyn childhood, I taught myself how to read even before I started school. Here's how it happened—

There was a rope-skipping game in which boys and girls, as they skipped, recited a set of nonsense syllables that went "ay-bee-see-dee—" and so on. Naturally, after a while, I knew it by heart.

I asked an older boy what it meant and he said it was the alphabet and wrote it out for me. I asked what it was for and he said that words were made out of it, wrote a few words, and told me what they were.

It seemed to me I detected a similarity in the sound of

27

the letter and the sound of parts of the words, so I asked him to tell me what each letter sounded like—and he told me that, too.

From then on, I began to find pleasure in working out the words on street signs, newspaper headlines, magazine covers, and so on. In all but the simplest cases, I tended to fail, but each time I succeeded and found myself pronouncing a word I knew, a pleasurable sense of accomplishment swept over me.

But this all took place when I was five years old and I do not remember the details to any great extent. One little event, however, stands out in living color. I will never forget it.

My mother and I were on an elevated railway heading southward through Brooklyn and I sat in the seat, my legs dangling, and studied the strip of white letters on black background that (although I didn't know it at the time) gave the destination of that particular train. It said C-O-N-E-Y I-S-L-A-N-D and I kept muttering to myself over and over "konnie issland—konnie issland—"

I was getting nowhere, but there was no point in asking my mother. Even if I could overcome my distinct feeling that asking someone else would spoil my fun, there was the fact that my mother had arrived not long before from Russia and that the very alphabet with which English was written was strange to her.

And then, somehow it occurred to me that we were going to Coney Island (pronounced k'nee-EYE-lin) and in a blinding flash of illumination I understood that what I was looking at was "koh-nee-eye-land."

To this day, nearly half a century later, I can remember the exact scene and my exact feeling—a sense of nearly unbearable delight of *understanding*. There are very few discoveries one makes nearly all by oneself, and here I had made two. First, I discovered there were such things as "silent letters," something that made it far easier for me to continue with my experimental working out of words. Second, I found that trains had signs telling their destinations, so that it became less miraculous that grown-ups knew where they were going.

I have known many pleasures since, many delightful stimulations of the various sense perceptions, many subtle bubbles of anticipation, and many lingering smiles of memory—but these are almost all trivial, evanescent, coming and going without affecting the world.

28

Learning—as much as possible by oneself, after puzzlement and effort—is something altogether different and can be compared to no other pleasure, for it spreads outward as far as the mind's eye can see. Learning is a joy that is mixed with a wide wonder as the universe seems to expand within the mental grasp, change and settle into a new form and color that it will never again lose. It's the "I've got it!" ecstasy (*eureka* in Greek) that, for a flashing moment, supersedes everything else; life, death, everything.

No, you can't do it *all* by yourself. There must be the kind of help that sets the foundation, gives you the springboard, guides you into position. Someone had to recite the alphabet to me, then tell me what it was, then write it for me, then tell me what each letter sounded like. There were even times when, puzzled, I broke down and asked someone to unriddle me something I could not manage. (I did that, for instance, with the word O-U-G-H-T, which I could, in no manner, manage to pronounce. And when I found some young man who could pronounce it for me, he could not explain what it meant.)

But then, no one lives in an intellectual vacuum. We observe, we are told, and even when we do not ask, information floods in.

We can react passively to the information flood; allow it to break over us while we ignore as much as we can with the most bovine of stolidity. And even then we cannot help but learn. Almost every child subjected to the stern rigor of grade school learns to read, write, and cipher, however much he may object to the process. And even before school, he learns to talk.

That's dull work, though—the royal road to knowledge —doing what you're told and absorbing, when you can resist no more, only what authority dictates.

Schools, unfortunately, are all too often exponents of this royal road to learning, with administrators and teachers incapable of doing anything but following the syllabus blindly, and with hardly anyone able to grasp the point that the most important fact a school can teach is—that school is not enough; that beyond the royal road to learning by way of the school's frowning authority lies the democratic road to learning by way of personal discovery.

Underlying this situation, at least in part, is the fact that the school system through all of history, since the first oldster taught the first youngster, has been undemocratic

29

in conception. Schooling is reserved, almost entirely, for one segment of the population—the young.

Learning is made to appear, in this way, a sometime thing, a temporary penance, imposed on the young and weak. What child is so dense as to fail to see that he must go to school, while grown-ups do not? What child is not quickly aware that one of the rewards of growing up is that at least he will be free to learn no more? Whatever a "dropout" may appear to us, to the child he seems someone who has managed to graduate to manhood in a hurry.

In how few places and on what few occasions is there ever an attempt to make it clear that learning is an integral part of the human condition; that to learn is to make use of that part of our body that is most peculiarly human; that to share in the accumulated store of knowledge gathered through time is the greatest of our human privileges. All else we share with all the rest of living species; our ability to learn is our own.

The true democracy of learning implies age-blindness. No one, of any age, who is anxious to learn in any field and on any subject, should be inhibited in that direction. He should be greeted with joy, seized upon with pleasure, and should receive whatever institutional aids are available.

This would be important if only because we are human; it has become essential because we are living in a society that, as far as age pattern is concerned, is different from anything mankind has known through its entire history.

Though the Bible speaks of "three score years and ten" as the lifetime of a man, it was a rare individual, indeed, who managed to live to be seventy prior to 1850. It could be done, yes, but half the children born in any year prior to 1850 were dead before the age of thirty-five, and in most places and at most times, before the age of twenty-five.

Now, in many parts of the world, and in the United States particularly, half of all those born can expect (assuming that our civilization does not deteriorate drastically in the coming decades) to live to be seventy. We live in a society crowding with men and women of middle age and beyond, and if the accent on a lowered birthrate succeeds, the percentage of the elderly within the population will continue to increase decade after decade for a considerable period of time.

And are we still to view education with the eyes of a culture in which the young made up the great majority of the population and in which the middle-aged did not really have to be considered because they died off with convenient speed?

Are we to condemn the potentially useful and creative coming-majority of the human race to the sterility of a social dictum that states, undemocratically, that learning is for one portion of the population only, the young? Are we to insist that old dogs can't learn new tricks to the point where the young dogs, growing old, take it for granted they can't learn new tricks and don't try, and then are we to use the fact that they don't try as proof that the assumption was correct all along? Shall the situation enforced by false doctrine be used to support the false doctrine?

By the democracy of learning, then, I mean the opportunity for schooling for anyone, at any age, in any subject of choice, with public encouragement and support.

And in saying this, I am not thinking of a school system that is merely expanded vertically to take in all age groups. No matter how you enlarge a school, you gain nothing if it remains merely a school.

Imagine a student sufficiently enamored of school to want nothing better than to stay for all the years he can. If he does so, and accepts school as nothing more than the royal road to learning by authority, he ends by being nothing but a dull specialist. Fie!

I myself, enjoying school, refused to budge until I had been awarded a Ph.D. plus a year of postdoctorate work in addition. And then when I finally announced my intention to leave, I made the smallest departure possible —from learning chemistry at Columbia University to teaching biochemistry at Boston University School of Medicine.

But all the years of application had merely plowed me into nearly parallel trenches of chemistry, trenches that steadily converged to a point some three inches before my nose. I didn't want that. I firmly believe that he knows nothing at all who knows only his specialty.

So I rebelled. Fortunately, either because of the lucky accident of my having experienced the delights of teaching myself to read before I ever went to school, or because the vicissitudes of life had cut me off from other and

31

lesser joys, I had continued to concentrate on self-education.

I have not ignored my specialty. I helped write a textbook of biochemistry (unsuccessful) for medical students and some other books on chemistry (reasonably successful) for laymen. However, I also wrote a three-volume book on physics, though I have taken exactly one course in physics in my life—in high school, at the age of fourteen. I have written several books on astronomy, though I have never taken any courses in that subject whatsoever.

I have written books on history and on literature, for that matter—and I have written fiction, too, but it's not generally supposed you need an education for that.

Let me give you a small example of just how even an extensive education can leave gaps, which can be permanent and damaging if one sticks closely to formal schooling, and nothing else, as a path to learning.

At no time—at *no* time—in all my long years of schooling, was I ever taught how to use a slide rule. Up to a certain point it was assumed that I didn't need to use a slide rule. After that point, it was assumed I already knew how to use a slide rule.

The first time I ever *saw* a slide rule was when I was a junior at college and picked one up in the office of the psychology professor. I fiddled with it and could make no sense out of it. I asked the professor what it was and he told me. I asked what it was used for and he answered that it was used to multiply, divide, and make other calculations.

Naturally I said, "Show me how it works," and just as naturally (for he was a busy man) he said, "I don't have time."

Once I had received that piece of instruction, I could hardly wait to buy one. Wait I had to, however, for it cost $16 and in those days $16 and $16,000 were alike in this respect—I had neither. It was not till after the war when my writing earnings were beginning to come in nicely that I could buy a slide rule for myself.

A book of instructions came with it. I began reading up to the point where I got the basic notion of the thing, and then I threw the instructions away lest I be tempted to return to it and spoil my fun. It took me quite a while of intensely amusing fiddling, but I learned how to work the thing.

Not all at once, either. I think it was not till two years

after I got the slide rule that, impelled by the exponential functions my research had mired me in, I learned how to use the log-log scales.

I daresay I didn't learn the slide rule as quickly or as well as I would have, had I received expert instruction, and to this day I may never have stumbled upon some little trick that might speed things up a little in certain calculations—but how dull that would have been. The fun and satisfaction of twiddling the slide rule till the right answers come tumbling out and you see what you're doing right rest far beyond the ABCs of instruction.

—Which, however, didn't prevent me from trying to instruct others (sorry, I can't help it). In 1965, I wrote a book on the subject—*An Easy Introduction to the Slide Rule*—and it came out just in time to be made useless by the arrival of the pocket computer. I was philosophic about that, however. I had written a story in 1950 that had predicted pocket computers and I couldn't object to being hoisted by the petard that I had myself foreseen.

Well, then, if schools aren't the full answer, what else do we need?

—Well, in 650 B.C. Ashurbanipal of Assyria collected bricks in his palace—stacks and stacks of bricks marked with fine cuneiform imprinting that held the gathered knowledge of 2500 years of culture in the Tigres-Euphrates Valley. Four centuries later, Ptolemy of Egypt (whom I mentioned at the beginning of this essay) began the process of accumulating the papyrus rolls that were to make up the largest library the world had seen up to that time.

Down to well into modern times that was the pattern of the storehouses of knowledge—they were the private property of kings and of great noblemen, and were available for use to a very few.

But now the smallest town can have a collection of books that rivals all but the greatest libraries of past ages, and, considering the advance of knowledge, contains in its least reference work wonders undreamed of by the great minds of the past.

So we must conclude that the most democratic place of learning in the world is the public library. It is there that we can find information on any subject, reading what we will, when we will, and how we will. And I look forward to the time when computerization will place in every home a terminal connected to some central library which will place, in facsimile, or on the television screen, the resources of human generations at the very fingertips of even the least of humanity.

But that, alas, was not in my time. In my time, the nearest thing to a distillation of human knowledge rested in the alphabetized multivolumes of an encyclopedia, where, for ready reference to any subject under the sun, you could reach from volume to volume without ever leaving a strategically placed chair.

I was an adolescent at the time the 14th edition of the *Encyclopaedia Britannica* first appeared and I couldn't help but note the advertisements therefor in the public prints. I don't think I ever wanted anything so much as a set of that encyclopedia. I fairly ached as the advertisements told me of the number of books on history—on science—on literature—on philosophy—even on sports —to which the *Britannica* was equivalent.

I stared at the sample illustrations and read the tiny bits of articles which they quoted, and I would look wishfully at my father.

But what was the use? Dreams are dreams, but if one is sensible, one does what one can with reality. So, after long and difficult negotiation, I managed to persuade my father to invest, at least, in a *World Almanac.* Heaven only knows where he got half a dollar from.

Finally, ten years later, the savings had piled up and the twenty-four volumes arrived in the house and we all stood around it in awe.

It was understood, after some discussion, that I had free access to them (my father, I think, was rather reluctant to have undue wear placed on the volumes by constant opening and closing) and I started reading.

No, I didn't read every word. But I began every article and continued reading till I lost interest. In a surprising number of cases I read the whole thing.

I was, however, only into the second volume when those well-known vicissitudes of life informed me that I would have to go to Philadelphia to work, and I never really returned home. As I remember, the last thing I did when I left, even after I said good-bye to my parents, was to jerk a thumb in the direction of the *Encyclopaedia* and say, "Now, I'll never know how the end comes out."

—But fear not. The time was to come when I finally got an *Encyclopaedia Britannica* of my own, and I discovered, at last, how it all came out at the end.* And it's one of my passports to joy to this day.

AFTERWORD

"The Democracy of Learning" appeared, as scheduled, in the fourth issue of Know. Alas, it was also the last issue, for with that fourth number, the magazine ceased publication. It just wasn't doing well enough.

And don't write to tell me that my article killed it. The decision was reached before the last number appeared.

---

* There's an index there.

4 THE MONSTERS WE HAVE LIVED WITH

FOREWORD

I am very fond of the editorial staff of TV Guide *(and have dedicated this book to them in consequence). I particularly like their habit of running "backgrounders" on special programs, for those are made to order for me.*

This article appeared in connection with a television special on some of the monster reports—the Abominable Snowman, Bigfoot, the Loch Ness monster. I always approach such popular beliefs in dramatic aspects of the Universe supported by shaky evidence (largely eyewitness evidence, which is virtually valueless) with complete skepticism.

Usually, this means that various True Believers, pained by what seems derision on my part, write me letters varying from hurt sorrow to annoyed rage. I even got letters of that sort arising out of this article on monsters, which seemed to me to have been very mild indeed.

Mankind has always lived with monsters. That fact dates back, no doubt, to the time when the early ancestors of man moved about in constant fear of the large predators about them. Fearful as the mammoths, sabertooths, and cave bears may have been, it is the essence of the human mind that still worse could be imagined.

The dread forces of nature were visualized as super-animals. The Scandinavians imagined the Sun and the Moon to be pursued forever by gigantic wolves, for instance. It was when these caught up with their prey that eclipses took place.

Relatively harmless animals could be magnified into terrors. The octopi and squids, with their writhing tentacles, were elaborated into the deadly Hydra, the many-headed snake destroyed by Hercules; into Medusa with her snaky hair and her glance that turned living things to stone; into Scylla with her six heads, whom Ulysses encountered.

Perhaps the most feared animal was the snake. Slithering unseen through the underbrush, it came upon its victim unawares. Its lidless eyes, its cold and malignant stare, its sudden strike, all served to terrorize human beings. Is it any wonder that the snake is so often used as the very principle of evil—as, for instance, in the tale of the Garden of Eden.

But imagination can improve even on the snake. Snakes can be imagined who kill not by a bite, but merely by a look, and this is the "basilisk" (from the Greek word meaning "little king").

Or else make the snake much larger, into what the Greeks called "Python," and it can represent the original chaos which had to be destroyed by a god before the orderly Universe could be created. It was Apollo who killed the Python in the early days of the Earth, according to the Greek myths, and who then established the oracle of Delphi on the spot.

Another Greek word for a large snake was "drakon," which has become our "dragon." To the snaky length of the dragon were added the thicker body and stubby legs of that other dread reptile, the crocodile. Now we have the monster Tiamat, which the Babylonian god Marduk had to destroy in order to organize the Universe.

Symbolize the burning bite of the venomous snake and you have the dragon breathing fire. Dramatize the swift and deadly strike of the snake and you have the dragon flying through the air.

Some monsters are, of course, animals that have been

misunderstood into beauty rather than horror. The one-horned rhinoceros may have contributed to the myth of the unicorn, the beautiful one-horned horse. And the horn of the mythical unicorn is exactly like the tooth of the real-life narwhal.

The ugly sea cow, with its flippered tail, raising half out of the sea and holding a newborn young to its breast in the human position, may have dazzled shortsighted sailors into telling tales of beautiful mermaids.

Throughout history, of course, man's greatest enemy was man, so it is not surprising that man himself served as the basis for some of the most fearful monsters—the giants and cannibalistic ogres of all sorts.

It may well be that the origin of such stories lies in the fact that various groups of human beings made technological advances in different directions and at different times. A tribe of warriors armed with stone axes, meeting an army of soldiers in bronze armor and carrying bronze-tipped spears, will be sent flying in short order with many casualties. The Stone Age survivors may well have the feeling that they have met an army of man-eating giants.

Thus, the primitive Israelite tribes, on first approaching Canaan and encountering walled cities and well-armed soldiers, felt the Canaanites to be a race of giants. Traces of that belief remain in the Bible.

Then, too, a high civilization may fall and those who follow forget the civilization and attribute its works to giants of one kind or another. The primitive Greeks, coming across the huge, thick walls that encircled the cities of the earlier, highly civilized Mycenaeans, imagined those walls to have been built by giant "Cyclops."

Such Cyclops were later placed in Sicily (where Ulysses encountered them in the tales told in the *Odyssey*) and were supposed to have but one eye. They may have been sky gods and the single eye may represent the Sun in heaven. It may also have arisen from the fact that elephants roamed Sicily in prehuman times. The skull of such an elephant, occasionally found, would show large nasal openings in front which might be interpreted as the single eye of a giant.

There can be giants in ways other than physical. Thus, medieval Englishmen had no notion of how or why the huge monoliths of Stonehenge had been erected. They blamed it on Merlin's magic. He caused the stones to fly through the air and land in place. (The Greeks also had

tales of musicians who played so beautifully that, capti-
vated by the sweet strains, rocks moved into place and
built a wall of their own accord.)

But as man's knowledge of the world expanded, the
room available for the dread or beautiful monsters he had
invented shrank, and belief in them faded. Large animals
were discovered—giant whales, moose, Komodo lizards,
okapis, giant squids, and so on. These were, however,
merely animals and lacking in the super-terror our minds
had created.

What is left then?

The giant snakes and dragons that once fought with the
gods and terrorized mankind have shrunk to a possible
sea serpent reported to be cowering at the bottom of Loch
Ness.

The giants, the ogres, the monstrous one-eyed cannibals
that towered over our puny race of mortals have diminished to mysterious creatures that leave footprints among
the snows of the upper reaches of Mount Everest, or show
their misty shapes fugitively in the depths of our shriveling
forests.

Even if these exist (which is doubtful) what a puny
remnant they represent of the glorious hordes man's mind
and imagination have created.

5 THE FOSSIL FUELS

FOREWORD

This article, and the one that follows, are largely about chemistry.

Through all the ages during which complex life has existed on Earth, occasional organisms died under conditions that did not allow them to decay. They fell into shallow water, into swamps, bogs, or mud, where the oxygen supply was short. Bacteria, which would ordinarily decompose the organisms, used up what oxygen there was in an early stage of the process and then also died.

The dead organisms remained undecomposed and the mud accumulated over them. The carbonate and phosphate atom groupings within hard parts such as bones and shells were replaced by silicate atom groupings bit by bit. The silicates are what make up the rocky crust of the Earth, so the bones and shells very slowly turned to a kind of rock while retaining all the structural details they had originally possessed. The mud, or sediment, which accumulated above them packed down harder and harder under its own weight and became "sedimentary rock."

All over the world, stony remnants of ancient life are found here and there in sedimentary rock. These remnants give us almost all the information we have concerning the details of the slow development of past life.

We call these rocky remains of ancient life "fossils." This is from a Latin word meaning "to dig" and was originally used to describe anything dug out of the ground. It has now come to be used specifically for the remains of ancient life. Dinosaur bones, turned to rock, are the fossils most familiar to the general public.

Not all organisms turn to stone with the ages, however. Plant life often does not.

About 345 million years ago, the Earth was mild in temperatures. It was not a mountain-building era and the continents were low-lying swampy plains, while shallow inlets of the ocean invaded the land here and there. The humid warmth encouraged vast growths of primitive trees.

When these died, they eventually fell into the shallow water, which was increasingly choked by dead plant organisms. The process of decay began and then stopped when the oxygen was used up. Other breakdown processes, not involving oxygen, took place in the choked swamps.

Plant tissue, it seems, and wood particularly, is made up of complex atom groups (or "molecules") containing atoms of carbon, hydrogen, and oxygen. Other types of atoms make up only a couple of percent of the whole.

The carbon and hydrogen atoms can combine with more oxygen than is present in the molecules. That is why wood burns; in doing so, it combines with additional oxygen in the air to form carbon dioxide and water.

As the plant tissue lay buried in water, portions of its molecules broke away. The parts that broke away were rich in hydrogen and oxygen atoms and poor in carbon atoms. These parts escaped as liquids or gases, leaving behind wood that possessed a larger and larger proportion of carbon atoms.

The waterlogged wood became "peat," in which accumulations of carbon atoms, combined with each other, made up about a third of the dry material, while the rest was still made up of carbon, hydrogen, and oxygen atoms in combination. Carbon atoms, in a mass, are black, so peat is usually brownish in color.

If the peat continues to remain in the water, mud and sediment will overlay it. Under the weight of the accumulating material, the peat is compressed. Its substance continues to break up and to grow richer and richer in carbon, so that its color becomes blacker and blacker.

The black substance, rich in carbon, we call "coal" and it comes in various grades, depending upon how long it has lain under pressure, and how much pressure it has been subjected to. The stage beyond peat is "lignite," which, when dry, is about 60 percent carbon.

Beyond that is "bituminous coal," which is some 85 percent carbon. Finally, in areas where particularly high pressures were applied because of folding of the Earth's crust, "anthracite coal" formed. This is 95 percent carbon.

In the waterlogged days of 345 million years ago, the swamps filled with whole forests of dead trees. These were slowly buried and huge seams of coal were formed, covered over with sedimentary rock. Sometimes several seams of coal alternate with sedimentary rock.

So vast is the quantity of wood turned into coal in this period of the Earth's history that the time (lasting for about 65 million years) is called the "Carboniferous Age," from Latin words for "coal-producing."

There are some 4.5 trillion tons of bituminous and anthracite coal in the ground, almost all of it formed from the forests of the Carboniferous Age. Add to that sizable quantities of lignite and peat. (Peat is still forming today in various boggy sections of the world.)

Coal is found wherever the large swamps of the Carboniferous Age existed and is therefore widespread. The swamps were particularly extensive in the land areas where now the United States exists. About one-third of all the coal in the world lies under American soil. However, Western Europe, the Soviet Union, and China also possess large reserves.

Coal, in all its forms, will burn, and it is therefore an example of a "fossil fuel." It doesn't necessarily burn in quite the same way wood does, however. The visible flame and smoke produced by burning wood is caused by the combination of oxygen from the air with hydrogen-containing molecules breaking away from the solid wood. The less the hydrogen-containing molecules are present in coal, the less coal flames and smokes when it burns. Anthracite coal, which is almost entirely carbon, merely smolders slowly, releasing its heat steadily over a long period.

(Wood can be heated to form a black high-carbon residue which will smolder slowly without flame or smoke. This is called "charcoal." Bituminous coal can be heated to drive off what hydrogen-rich fragments it possesses and what is left behind is "coke.")

In ancient and medieval times coal was sometimes used as fuel, but ordinary wood was much easier to obtain and much easier to set on fire. As late as 1760, wood was virtually the sole fuel used by mankind.

By then, though, the forests had disappeared from much of Europe, particularly from Great Britain. The growing industrialization of Great Britain, combined with the disappearance of its forest reserves, made it necessary to use coal.

It was the burning of coal that powered the Industrial Revolution of Great Britain, and then of the rest of the world. It was the fact that coal was found in so widespread a fashion over the lands of the world that made it possible for the Industrial Revolution to spread as it did.

By 1900, some 96 percent of the world's energy was derived from burning coal. Other fuels then came into use, so that coal's percentage of the total dropped, but the overall quantity used continued to rise.

At the present time, the world is burning about 3 billion tons of coal per year. The world's industry, as far as it depends upon coal, is consuming the energy capital that was very slowly stored out of the dying forests of the distant past. These forests built up their tissues by using the energy of Sunlight, so that, in burning coal, we are using some of the Sunlight that poured down on Earth hundreds of millions of years ago.

To be sure, peat and lignite and coal are still forming today, but at a very slow rate. We are burning it all at least 50,000 times as fast as it is forming. Coal is not, therefore, a renewable fuel. Eventually it will be gone and we will not be able to wait for more to form.

At the rate at which we are presently using coal, there remains enough in the ground, if all could be used, to supply us for 1500 years. However, much of the coal is too well-buried or exists in seams too thin for economical recovery. It might be safer to suppose that the world's practically available supplies will last us, at the present rate of use, for not more than 500 years, provided other fuels now being used also endure.

The greatest disadvantages involved in the use of coal are the dangers of digging it out of the ground and the difficulty of shipping it in vast quantities from the place where it is found to the place where it must be used.

These disadvantages arise out of the fact that coal is a solid. There are other fossil fuels, however, that are not. The chief of these is petroleum, which is a liquid. Petroleum differs from coal, chemically, in that it is composed not of carbon atoms chiefly, but of carbon atoms in combination with hydrogen.

Petroleum, like coal, is found in sedimentary rocks, and was probably formed from long-dead living organisms. The rocks in which it is found are almost always of ocean origin and the petroleum-forming organisms must have been ocean creatures rather than trees.

Instead of originating in accumulating woody matter, petroleum may be the product of the accumulating fatty matter of ocean organisms such as plankton, the myriads of single-celled creatures that float in the surface layers of the ocean.

The fat of living organisms consists of atom combinations that are chiefly made up of carbon and hydrogen atoms. It does not take much in the way of chemical change to turn that into petroleum. It is only necessary that the organisms settle down into the ooze underlying shallow arms of the ocean under conditions of oxygen shortage. Instead of decomposing and decaying, the fat accumulates, is trapped under further layers of ooze, undergoes minor rearrangements of atoms, and finally is petroleum.

Petroleum is lighter than water and, being liquid, tends to ooze upward through the porous rock that covers it. There are regions on Earth where some reaches the surface and the ancients spoke of pitch, bitumen, or asphalt. In ancient and medieval times, such petroleum seepages were more often looked on as medicines rather than as fuels.

Of course, the surface seepages are in very minor quantities. Petroleum stores, however, are sometimes overlain with nonporous rock. The petroleum seeping upward reaches that rock and then remains below it in a slowly accumulating pool. If a hole can be drilled through the rock overhead, the petroleum can move up through the hole. Sometimes the pressure on the pool is so great that the petroleum gushes high into the air. The first successful drilling was carried through in 1859 in Titusville, Pennsylvania, by Edwin Drake.

If one found the right spot (and prospectors eventually learned to recognize the kind of geologic formations that made it likely for a pool of trapped petroleum to exist underground) then it was easy to bring up the liquid material. It was much easier to do that than to send men underground to chip out chunks of solid coal. Once the petroleum was obtained, it could be moved overland through pipes, rather than in freight trains that had to be laboriously loaded and unloaded, as was the case with coal.

The convenience of obtaining and transporting petroleum encouraged its use. The petroleum could be distilled into separate fractions, each made up of molecules of a particular size. The smaller the molecules, the easier it was to evaporate the fraction.

Through the latter half of the nineteenth century, the most important fraction of petroleum was "kerosene," made up of middle-sized molecules that did not easily evaporate. Kerosene was used in lamps to give light.

Toward the end of the nineteenth century, however, engines were developed which were powered by the explosions of mixtures of air and inflammable vapors within their cylinders. The most convenient inflammable vapor was that derived from "gasoline," a petroleum fraction made up of small molecules and one that therefore vaporized easily.

Such "internal combustion engines" are more compact than earlier steam engines and can be made to start at a moment's notice, whereas steam engines require a waiting period while the water reserve warms to the boiling point.

As automobiles, trucks, buses, and aircraft of all sorts came into use, each with internal combustion engines, the demand for petroleum zoomed upward. Houses began to be heated by burning fuel oil rather than coal. Ships began to use oil; electricity began to be formed from the energy of burning oil.

In 1900, the energy derived from burning petroleum was only 4 percent that of coal. After World War II, the energy derived from burning the various fractions of petroleum exceeded that of coal, and petroleum is now the chief fuel powering the world's technology.

The greater convenience of petroleum as compared with coal is, however, balanced by the fact that petroleum exists on Earth in far smaller quantities than coal does. (This is not surprising, since the fatty substances from which petroleum was formed are far less common on Earth than the woody substances from which coal was formed.)

The total quantity of petroleum now thought to exist on Earth is about 14 trillion gallons. In weight that is only one-ninth as much as the total existing quantity of coal and, at the present moment, petroleum is being used up much more quickly. At the present rate of use, the world's supply of petroleum may last for only thirty years or so.

There is another complication in the fact that petroleum is not nearly so evenly distributed as coal is. The major consumers of energy have enough local coal to keep going but are, however, seriously short of petroleum. The United States has 10 percent of the total petroleum reserves of the world in its own territory, and has been a major producer for decades. It still is, but its enormous consumption of petroleum products is now making it an oil importer, so that it is increasingly dependent on foreign nations for this vital resource. The Soviet Union has about as much petroleum as the United States, but it uses less, so it can be an exporter.

Nearly three-fifths of all known petroleum reserves on Earth is to be found in the territory of the various Arabic-speaking countries. Kuwait, for instance, which is a small nation at the head of the Persian Gulf, with an area only three-fourths that of Massachusetts and a population of about half a million, possesses about one-fifth of all the known petroleum reserves in the world.

The political problems this creates are already becoming crucial.

Some of the molecules that occur in petroleum are so small that they make up substances which are gases at ordinary temperature. In many ways, gases are even more convenient than liquids, when it comes to obtaining, transporting, and using. The so-called "natural gas," found in conjunction with some petroleum sources, has grown especially popular since World War II. In the United States, particularly, natural gas is now supplying over a third of the energy being used.

One advantage of natural gas is that of all the fossil

fuels it is cleanest. Both coal, although made up mostly of carbon atoms, and petroleum, although made up mostly of carbon and hydrogen atoms, have a small percentage of other atoms. Some of these other atoms produce substances, when the fuel is burned, that are unpleasant in one way or another even though they may be present in only small quantities.

Sulfur atoms are particularly villainous in this respect, producing sulfur dioxide on burning, which, in combination with water, produces acids. The sulfur products irritate throats and lungs and damage the environment in many ways. They are an important component of smog and are an example of the kind of "air pollution" that is becoming a major problem in the world today and particularly in the cities of the industrial nations.

To avoid this, coal or petroleum must be found which is free of pollution-prone substances (which cuts down the quantity available for use) or must be made free (which raises the expense).

The small molecules of natural gas are made up *only* of carbon and hydrogen and therefore burn without air pollution. However, to balance that, natural gas is, unfortunately, less common even than petroleum. There are 720 trillion cubic feet of natural gas in the world, it is estimated, and nearly two-fifths of this quantity is to be found in the United States. At the present rate of use, it will last only twenty years or less.

Is all this too pessimistic? After all, in recent decades, new reserves of fuels have been discovered. The huge petroleum reserves of the Middle East were unsuspected several decades ago and lately we have discovered petroleum reserves in Alaska. Might there not be vast petroleum reserves still waiting to be discovered on the continental shelves, for instance?

Yes, there might; but the skill of prospectors has been increasing and we are running out of places to look. New findings are declining and this, indeed, is inevitable. The Earth is only of finite size and can hold only so much petroleum.

Of course, we might turn to petroleum supplies less rich than the pools trapped under impermeable rocks. There are places where petroleum simply seeped up into the porous rocks and remained there near the Earth's surface. We could exploit so-called "oil shale" or "tar sands." Pe-

47

troleum from such places is harder to obtain and it will therefore be more expensive—and such sources, too, are limited.

Liquid and gaseous fuels can be formed out of coal, which is, as yet, not in imminent danger of being used up. That would mean, though, that coal would have to be mined at more than double its present rate and transported in that much higher quantity over badly strained rail networks.

If coal became the source of all the various types of fuel we use, the reasonably available supply, even at present rates, might last us only 200 years.

Might we not become more efficient in the use of the fossil fuels of every kind? At the present time, at least 60 percent of the energy we get by burning them is wasted in one way or another and just escapes as unused heat.

True! But even if we increased the efficiency to 100 percent (and we can go no further than that) we will stretch out our fossil fuel supply to only a little more than twice what we can now count upon.

Can we not simply pull in our horns and use less energy? Can't we recede from electric shavers and electric toothbrushes and electric typewriters and go back to hand devices? Can't we give up electric air conditioners and sweat a little? That's certainly possible, but there are pressures against it.

Not only is there the fact that people hate to give up their comforts, but a large part of the world, still undeveloped at present, longs to increase its standard of living, and this can be done only by their using energy at a much greater rate. Then, too, the world population keeps going up, and as long as that happens, total energy use may well increase even though each individual uses less.

There is another thing to consider. Burning any of the fossil fuels produces carbon dioxide. This in itself is not a terribly dangerous substance; it is always present in our bodies and is present in the atmosphere to an extent of 0.03 percent. It does us no harm in that percentage and would do us no harm if the quantity were doubled or tripled.

Carbon dioxide retains heat, however. Sunlight passes through it, but the warmed Earth emits infrared radiation which carbon dioxide absorbs and retains. The Earth therefore remains somewhat warmer than it would be if there were no carbon dioxide in the atmosphere. If the

48

carbon dioxide content goes up even slightly—not enough to harm us directly—the Earth will become slightly warmer.

No matter how one cleans up fossil fuels to prevent ordinary air pollution, the production of carbon dioxide will continue. If we imagine all the coal and petroleum that now exists in the Earth to be burned, the amount of carbon dioxide produced would be ten times as much as is now present in the atmosphere.

Not all of it would stay in the atmosphere, of course; some would go to support a more luxuriant plant growth and some would dissolve in the sea. Enough would remain in the air, however, to warm the Earth a few degrees, and though it would not harm us directly, it would melt the ice caps and produce many changes that might prove catastrophic.

On the whole, then, there is a limit to how much we might want to burn fossil fuels and we might even want to stop the process while there are still considerable reserves remaining in the ground.

We can summarize thus:

The fossil fuels supplied by the plant and animal world in ages past have done mankind good service. Coal supported man's advancing technology through the nineteenth century and early twentieth century. Petroleum is supporting man's advancing technology through the middle twentieth century and, in all likelihood, will continue to do so through what remains of it.

By the twenty-first century, however, even if we achieve population stability, increase the efficiency with which we use fossil fuels, discover new reserves, and economize on our energy demands, we will still be forced to turn from fossil fuels and exercise other energy options, if we expect to continue to support our complex technology and our comfortable affluence.

Fortunately, such alternate energy options exist, and if we use our ingenuity and avoid catastrophes of other sorts, the energy crisis will not destroy us.

Eventually, mankind will look back across the centuries and say, perhaps, that the fossil fuels, though they virtually disappeared in the first two centuries of industrialization, kept us going till we learned to use more permanent energy sources—something we might not have learned without the schooling of those two fossil-fuel centuries.

AFTERWORD

In the issue of National Wildlife *in which the preceding article first appeared, there was an editorial entitled "Energy Crisis: How Big? How Bad? How Soon?"*

The answer came remarkably quickly, for three months after this issue of the magazine was published there came a brief new war in the Middle East and that was followed by the Arab oil boycott. That forever put an end to our illusions concerning our independence of the rest of the world.

Incidentally, although the preceding article was written eight months before the boycott, not a word of it had to be changed because of that. The boycott revealed nothing that any rational human being did not already know.

6 A DROP OF WATER

The next article appeared originally in a magazine named
Poolife, *which is published by a company that manufac-*
tures chemicals for the disinfecting of swimming pools.

In my salad days as a writer of miscellaneous articles,
I hadn't the slightest idea of the vast number of magazines
that existed and flourished. The ones most people are
aware of are those that hog the front-row space in the
magazine displays, but there are myriads of other maga-
zines hungry for material and very often interestingly spe-
cialized.

Would you have thought there were enough owners of
swimming pools to support a magazine? Well, there are,
and the magazine has a sturdy circulation and flourishes
sufficiently well to be able to pay handsomely for articles.

A drop of water? That's an easy thing to find on Earth,
since our planet possesses something like 28 trillion trillion

drops of it—or 28,000,000,000,000,000,000,000,000,000,000 drops, if you want the number written out in full.

Why so much? It's the result of the recipe of the Universe. The best astronomical estimate is that over 90 percent of all the atoms in the Universe are hydrogen atoms, the simplest of all. Out of every 2000 atoms in the Universe, about 1860 are hydrogen, and 138 are helium, the second-simplest atom. There is, in addition, 1 atom of oxygen. That leaves one last atom which might be any one of the remaining hundred or so known elements.

Helium atoms are completely standoffish. They do not combine either with themselves or with atoms of any other element. Not so with hydrogen and oxygen. If the temperature is reasonably low, they will combine to form atom groups or, as they are called, "molecules." Two hydrogen atoms will combine with each other to form a hydrogen molecule, which can be written in chemical shorthand as H_2. Two hydrogen atoms will also combine with an oxygen atom to form a water molecule (H_2O). The cool matter of the Universe, the dust and gas that is not too near a star, is therefore made up chiefly of hydrogen molecules, helium atoms, and water molecules in that order.

When planets form out of cold matter, they collect at their core those relatively rare atoms that form solid matter such as metals and rock. Around that core, the hydrogen, helium, and water collect. If the planet is large and cold, a great deal of these common materials collect and you have a giant world like Jupiter. If the planet is small or hot or both, there isn't enough gravity to hold the light hydrogen and helium or the almost-as-light water. The planet is then solid matter and nothing more, like the Moon and Mercury.

If the planet is middle-sized, it is still not large enough to hold the hydrogen or helium, but it can hold a lot of the water, enough water to form an ocean. Earth is an example of that.

Earth's ocean is the key to its life. The important molecules of living tissue are large and complicated, each containing thousands or even millions of atoms in precise arrangement. The properties of life depend on the intricate and rapid interchange of atoms, atom groups, and sometimes even parts of atoms, among these complicated molecules.

Such an interchange cannot take place in solid matter,

because there all the molecules are more or less fixed in place and interchanges can take place only very slowly. Nor can it take place in gaseous matter because there molecules are far apart from each other, so that interchange can't be fast enough.

In liquid matter, molecules are close enough together to be in contact, and yet they roll and slide over each other freely. That's where interchange takes place rapidly enough.

What's more, it isn't just interchange among the atoms making up the water molecule that we deal with in the ocean. Water has the ability to cancel out some of the chemical forces that hold molecules together. Dry substances separate into single molecules in the presence of water, sometimes into molecular fragments. In other words, water dissolves many substances. It is, in fact, a better all-round solvent than any other common liquid.

Even where water does not actually dissolve a particular substance, there is enough attraction between water molecules and other kinds of molecules to cause the water to spread out over the surface of most substances and cling to it. That is why water wets objects. (Wax is one substance that does not attract water molecules, so that a candle, for instance, can be dipped into water and come up dry.)

Because of water's ability to wet and dissolve substances, the ocean is full of substances other than water; it is 3.5 percent dissolved material. Most of this dissolved material is salt, but smaller quantities of every element on Earth (in one combination or another) are to be found in the ocean. This means that in the ocean there is a wide variety of atoms combining, interchanging, and breaking apart.

In Earth's early years of existence, the energy of Sunlight tended to force atoms to combine into larger and larger molecules until, finally, molecules were formed which were complex and intricate enough to possess some of the properties we associate with life. Life formed in the ocean, then, over three billion years ago.

It was not till 400 million years ago, though, that life managed to push out of the ocean onto the dry land. And even then, living things carried some of the ocean with them. We still have the ocean in our own bodies. Our bloodstream is very much like a current of ocean water, complete with dissolved materials, and carrying along with

53

it the complex molecules of life and whole cells along with it. The cells of our body are watery, too. The human body as a whole is 60 percent water, and the more active organs are up to 80 percent water.

Nor is the dry land actually dry. Only 30 percent of the Earth's surface is land and it is permeated with water, thanks to the presence of Earth's vast 70-percent-encompassing ocean.

The water molecules of the ocean evaporate in the Sun's heat, while the solid substances dissolved in the ocean remain behind. If this process were to continue forever, the ocean would dry up, but, of course, it can't. Some of the water vapor is always returning to the ocean and at any given temperature the atmosphere can only hold so much water vapor.

What's more, cold air can hold considerably less water vapor than warm air can. If warm air cools, either because it lifts to a higher altitude or because it moves out of the tropics and toward the polar region, it may no longer hold all the vapor it contains and some of it will turn to liquid water and fall as rain. (If the temperature is low enough, it will fall as snow.) In the course of a year, some 350,-000 cubic kilometers (85,000 cubic miles) of water is evaporated from the ocean and then falls again as rain or snow.

Not all the rain or snow falls directly on the ocean. Some water vapor is blown over the land areas and the rain falls there. That rain eventually returns to the ocean, too, but not necessarily right away.

Rain falling on land soaks into the ground until it reaches a layer of impermeable rocks and can go no farther. The water that accumulates over that rock has a surface that is called the "water table." It spills over into the portions of the land surface that are lower than surrounding areas to form ponds and lakes. These overflow, too, and pour downhill as streams and rivers.

Since it is only the water portion of the ocean that evaporates, it is almost pure water that falls as rain or snow. The lakes, ponds, and rivers on land therefore all represent "fresh water" with very little dissolved material in it, as opposed to the "salt water" of the ocean.

As the fresh water flows through or over the soil and makes its way back to the ocean, it does pick up a small amount of dissolved matter. The exact nature of this matter depends on the makeup of the surrounding soil. Very often the dissolved matter is freely soluble and doesn't tend to come out of solution again. Such water is "soft water." Where the dissolved substances are compounds of calcium and magnesium, for instance, there is a tendency for some of it to cake out and corrode machinery, or to combine with soap and reduce its cleaning power. This is "hard water."

Fresh water contains life, too, though not as richly as does the salt water of the ocean. Life makes use of the dissolved solids in water, and since fresh water has much less of these solids, it can support only smaller quantities of life.

Plants that grow on "dry" land are also supported by water. Their root systems gather water from the land immediately below them. Where there is much rainfall, plant growth can be luxurious indeed, but even where there is little rainfall, some plants can make it. In dry areas, plants develop intricate root systems that can reach down to the very low water table, and they are adapted to store whatever they get very efficiently. Even the driest deserts have occasional areas where the water table is high enough to support plant life and there we have oases.

As for animals, they get their water from the moist tissues of the plants they eat. In the drier areas, animals are adapted to strict water conservation, so that what they get from their food is all they need. Most animals, however, live in places where free water is easily available. They have not adapted to strict water conservation and they must drink water to supplement the quantity they find in their food. This is true for human beings, for instance.

Water remains an attraction of land life. Bodies of water, particularly the ocean, are richer in life than the land is, and land organisms are drawn to the ocean as a food supply. In every group of land organisms there are examples of life forms that have adapted themselves for a return to the ocean—in some cases to the point where they cannot live on land anymore.

Consider the whales and dolphins. They are mammals; they breathe air and bear live young. Their ancestors were land-dwelling organisms, but they themselves can no longer return to land. They can live only in water.

Even land organisms which remain primarily land-living may haunt the water for food. Fish serve as food for countless species of mammals and birds. Man, too, is finding the ocean a more and more important source of food.

Nor is water attractive only for its food supply. It is a natural place where one can have fun and relax. Some animals seem to find this to be so, and certainly human beings do.

Water, the age-old cradle of life, offers a few things that the dry land cannot offer. For one thing, water is a much denser medium than air and its buoyancy virtually neutralizes the pull of gravity. Living organisms never experienced the savage, undiluted, relentless pull of gravity till they emerged onto land, where only the insignificant buoyancy of air existed to counter it. Land life had to develop strong limbs, therefore, with which to fight that eternal pull.

There is a certain age-old relief, then, to move into water and feel that pull vanish. Floating and wading are fun for that reason alone.

Because water is a liquid and because within it gravitational pull is not strong, life in water can move three-dimensionally—not only east, west, south, and north, but up and down, too. Yes, on land, birds can experience the same three-dimensional movement through air, but air is so thin a medium that only light organisms can fly. No flying organism is, or probably ever has been, as much as 25 kilograms (55 pounds) in weight. Nor is flying an easy occupation; it requires the expenditure of energy at a great rate.

Water, however, a much denser medium, holds one up and allows motion to be, not fast perhaps, but lazy and graceful. What's more, it can hold up virtually any weight. The blue whale, which can have a weight of more than 150 tons, and whose tongue alone is heavier than an elephant, is easily supported by water. Yet it is not only larger than any land organism today, but it may, indeed, be the largest organism that has ever lived.

It is no wonder, then, that human beings find pleasure in plunging into water, particularly fresh water. In warm weather, the plunge is cooling and invigorating; it tends to wash off accumulated grime (sea water will do the same, but then fresh water must be used to wash off the dried salt caking our skins); it supplies an environment in which we can have the fun of forgetting about gravity, so that we can float and dive.

Humanity has this advantage over other species of animals, moreover—it possesses a technology that is capable of producing an artificial pond at some convenient location (the back-yard pool) which can be reserved for the private use of selected human beings only. Other species of organisms are, of course, kept out. Otters, pelicans, even goldfish, are not welcome in a swimming pool.

However, water means life, and if sizable organisms can be kept out of the pool with ease, microscopic ones are much more difficult to handle. A pool automatically gathers bacteria and algae unless chemical defenses against them are used.

All the careful art involved in constructing and maintaining a pool gives evidence, then, of the powerful pull of the ancestral watery environment on us, even after hundreds of millions of years of life out of water.

AFTERWORD

Ethical problems sometimes arise in connection with miscellaneous writing such as my own. Ought I, for instance, tell the publishers in the case of the preceding article that I have never owned a swimming pool and that I can't even swim? Are pool owning and swimming a prerequisite for writing such an article?—Obviously not.

Still, that made me feel guilty enough to be certain that I threw in that final, oblique reference to disinfecting pools. It fit the article well enough, so why not shake the hand that feeds me?

B

THE
PRESENT

7 SMART, BUT NOT SMART ENOUGH

FOREWORD

The next article, like all those I write which accept the fact of evolution, elicited numerous reactions, some sorrowful, some vitriolic. It seems quite obvious that many people flinch at any suggestion that the marvelous complexity of human beings was formed by a slow and incredibly intricate process stretching over the entire fifteen-billion-year lifetime of the Universe. They seem to think it makes more sense to suppose human beings to be molded out of dirt in an instant by a primitive super-being much given to petulance and wrath.

My own reaction to letters objecting to my evolutionary views was to handle the subject of this article at greater length in one of my F & SF essays. It then became "Look Long Upon a Monkey," which was eventually included in my essay collection Of Matters Great and Small *(Doubleday, 1975).*

If a hundred million monkeys were to pound at type-writers long enough, they would eventually write all the books in the British Museum.

Isn't that an impressive statement? —But why monkeys?

The oft-repeated remark about the hundred million monkeys is an expression of the power of random events, given sufficient time; and an important power it is, since it is what produced man out of the chemical soup of the primordial ocean. But randomness doesn't require monkeys. Hook the typewriters to any inanimate device designed to produce random depressions of the keys and the principle would be served just as well. Why monkeys, then?

Because monkeys are smart, but not smart enough. You can easily imagine monkeys being smart enough to pound typewriter keys in imitation of men, but you can't imagine them smart enough to type them in any way other than at random.

Smart, but not smart enough— Monkeys amuse people because they are smart enough to be like people in some ways; and they bother people for the same reason. Why were these grotesque imitations of ourselves created? Were they practice attempts before the creation of man? (Mark Twain said it was the other way around.)

Or are monkeys relatives of ourselves and did we have common ancestors many millions of years ago, as Charles Darwin suggested? *That* thought of relationship bothered many people. It still does. It became important to many people to insist on some great and fundamental difference between man on the one hand and the entire monkey tribe (the "primates") on the other; a difference that could not be transcended, so that there would be no question of a relationship.

Monkeys alone might not have been so bad. They had tails like other animals instead of the smooth rear end characteristic of man, and that was enough to rule them out. (The Barbary ape is an exception; it is a tailless monkey, native to northwestern Africa.)

Then, as Europeans explored Africa and the East Indies, the true apes were discovered. These were larger than monkeys, as tailless as man, and far closer to man, in their caricatured appearance, than monkeys were.

The discovery came surprisingly late. It was not till 1698 that the first chimpanzee was brought to Europe, not till 1776 that the first orangutan was. The gorilla was not properly described till 1847, just twelve years before Darwin's book was published.

Apes bring up the question of relationship far more strongly than tailed monkeys do. Even the most casual study shows that their resemblance to man is uncanny. If one dissects apes, it turns out that in the most intimate details of their inner workings, the resemblance remains uncanny. Modern techniques of testing the fine structure of protein molecules show that the chimpanzee's proteins are closer to man's than to any other species. The chimpanzee is closer to man, biochemically, than it is to the gorilla.

The most important physical difference between man and the chimpanzee lies in the size of the brain. The human brain is four times as large as the chimpanzee brain. Except for size, though, they're alike. Might not the difference, then, be one of degree only?

Not necessarily. When a difference in degree becomes large enough, it can become a difference in kind. The human brain, four times as large as the chimpanzee's, becomes capable of matters infinitely beyond the poor ape. Even if we were to grant that the chimpanzee is more intelligent than any other animal, surely its intelligence is as nothing compared to that of a man.

To those who will not endure the thought of any relationship between man and other primates, the mental difference remains, even if the physical difference vanishes. It can be argued that the gap in intelligence between man and even the most intelligent of other animals is too great to be bridged.

For instance, man can talk. Other animals (even as low in the scale of life as bees) can signal, in one way or another, primitive emotions and desires—"I'm scared"; "I'm hungry"; "Go away"; "Let's make love"; "I'm lost." Only man, however, can develop a complicated set of modulated sound symbols to express the most delicate shades of meanings and the most esoteric abstractions.

Many attempts have been made to teach young chimpanzees to talk. They have all failed. Chimpanzees brought up with human babies develop more quickly and outstrip their child companions until the stage at which children learn to talk. Then the baby chimpanzee is outdistanced forever. It is smart, but not smart enough.

Men can talk because they are capable of handling the muscles of their tongues, throat, lips, and so on with the greatest delicacy. The ability to do so is governed by a part of the brain called "Broca's convolution." Damage

63

that section and a human being can no longer speak or understand speech, but he can still communicate by gesture.

Chimpanzees lack the equivalent of Broca's convolution, but in the wild they communicate by gesture, too. It occurred to Beatrice and Allen Gardner at the University of Nevada, in 1966, to try to teach a deaf-and-dumb language to a one-and-a-half-year-old female chimpanzee they named Washoe. They were amazed at the results. Washoe learned dozens of symbols, used them correctly, understood them easily, and made up new combinations of symbols which she also used appropriately.

Other chimpanzees were also taught. One chimpanzee was taught to manipulate magnetized counters bearing symbols, and learned to understand and construct sentences with all due attention to grammar and punctuation —the near-equivalent of reading and writing. Chimpanzees have been taught to manipulate those symbols by means of a kind of typewriter. Chimpanzees have taught symbols to other chimpanzees, and have also tried to teach them, without success, to pet cats.

So chimpanzees can communicate in the equivalent of a simple human language, and even in terms of abstractions, provided you place that language in terms of movements they are physiologically equipped to handle. The difference between the chimpanzee and man is one of degree only, after all—even in intelligence.

Can we teach chimpanzees enough to put them to work for us? Other animals have been tamed and put to work—dogs, horses, reindeer, llamas, yaks, oxen, camels, elephants, and so on. They aren't as intelligent as chimpanzees, of course, but surely intelligence is no disqualifier.

After all, human beings can also be trained to do enforced work. The slave, under the overseer's lash, is treated as no more than a more intelligent horse.

Why, then, not a chimpanzee as well? In my opinion, that won't work. Chimpanzees are too smart to settle down to dull, repetitious, straining labor in response to human signals, as horses and even elephants do. On the other hand, they are not smart enough to understand the futility of rebellion as human slaves do, or clever enough to make slavery as tolerable as possible, as human slaves sometimes do.

In short, chimpanzees are smart enough to do the work

—but not smart enough to be enslaved. —And maybe that *is* smart enough.

AFTERWORD

The most amusing objection to my series of TV Guide articles, incidentally, was from one innocent who complained that I got all my material out of encyclopedias. Apparently he thought I ought to make up all factual matter out of my head.

Bill Marsano of TV Guide, a prince of good fellows, wanted me to answer, but wouldn't listen to my attempts to explain that originality consisted in the organization and expression of facts, not the facts themselves. He said, "No, no, Isaac. Be funny!"

So I jotted down a note to the effect that I used two encyclopedias, taking words alternately from each, and they printed that.

8 RECIPE FOR AN OCEAN

FOREWORD

When this article first appeared in Natural History, *my title was changed by the addition of a single word. It became "Recipe for a Planetary Ocean." Title changes don't disturb me. For one thing I recognize the need of magazines to attract readers by titles, whereas I need only satisfy my own sense of fitness. For another, since my articles are generally going to appear in one or another of my collections, I know that I can restore my original title—as I have done in almost every case in this book. (Sometimes the magazine title turns out to be an improvement, in which case I quietly adopt it.)*

In this case, though, I must own up to a wild curiosity. Why add the adjective "planetary"? What else can an ocean be? It is planetary by definition.

Earth has an ocean. Its ball of solid material is covered by a film of moisture. At least it is a film by planetary standards, but it is up to 11 kilometers deep in some places, just the same. Land surface pokes up through the ocean in places, but the continents and islands make up only 30 percent of the Earth's surface.

Is this a common situation? May we expect any planet to have an ocean? Will it always be a water ocean as Earth's is?

Or is an ocean a rare thing?

To answer the question, let us consider what the requirements of an ocean are. First, it must be made of a substance that is liquid at the temperature and atmospheric pressure of the planet's surface. Second, it must be a substance that is cosmically common, so that enough of it will be found on a planet to form an ocean.

There are not many elements which are cosmically common. Only fifteen might be listed: hydrogen, helium, carbon, nitrogen, oxygen, neon, sodium, magnesium, aluminum, silicon, calcium, sulfur, argon, iron, and nickel. Any substance that is not either one of these fifteen elements, or a compound made up of two or more of these fifteen elements, cannot possibly be present on any planet in sufficient quantity to make up an ocean.

If we start with these fifteen elements, can we foretell what combinations will form?

To begin with, helium, neon, and argon will not enter into combinations with other elements, or among themselves, for that matter. They will remain uncombined elements.

Then, hydrogen is the predominant component of the cosmic mixture, making up an estimated 90 percent of all the atoms in the Universe. Hydrogen combines with other elements rather freely, but even if it did so with all the other material available, a great deal of it would be left over uncombined.

Of the other elements, carbon, nitrogen, oxygen, and sulfur will easily combine with the superabundant hydrogen to form, respectively, methane, ammonia, water, and hydrogen sulfide.

Oxygen would also tend to combine with silicon and the combination would itself tend to combine with the sodium, magnesium, aluminum, and calcium to form "silicates," the stuff of which the rocky crust of the Earth is made. Iron and nickel may also be found in the silicates, but the tendency for this is rather small, and an iron-nickel mixture (9 to 1 in proportion) is likely to remain uncombined.

We see, then, that the possible recipe for the manufacture of a planet will include the following as ingredients: hydrogen, helium, neon, argon, methane, ammonia, water, hydrogen sulfide, silicates, and iron-nickel.

We can divide these ten ingredients into three classes.

The first class includes hydrogen, helium, neon, and argon. These have boiling points below —170° C and are going to be gases under all but the most unusual conditions. They are not likely to be ocean-forming substances.

The second class includes silicates and iron-nickel. These have melting points above 1000° C. and are going to be solids under all but the most unusual conditions. They, too, are not likely to be ocean-forming substances.

That leaves us with the third class: methane, ammonia, water, and hydrogen sulfide. These are the only substances that might be liquid under reasonable conditions and that can be present in quantities sufficient to form an ocean.

Next, let's consider the conditions under which a planet can form (and the word "planet" is meant here to include smaller bodies, too, such as satellites and asteroids).

The chief variable in the formation of a planet is its distance from the central star. The planet can be forming relatively close to the star, or relatively far from it.

If the planet is forming close to the star, its temperature is going to be comparatively high and all the atoms and molecules coming together to form it will be moving comparatively rapidly.

Under these conditions the small, and therefore particularly nimble, atoms of helium and neon cannot be held by the gravitational field of the forming planet. Neither can the small two-atom molecules of hydrogen be held. Since hydrogen, helium, and neon, together, make up some 99 percent of all the atoms or molecules in the original mix, the planet, forming out of the scraps that are left, *cannot* be large.

If the planet is sufficiently close to the central star, or if it is particularly small, it can't even hold the somewhat heavier molecules of the third-class substances—the so-called "volatiles" (because even when they are liquid they can easily volatilize, or turn to gases). All that is left are silicates and nickel-iron, the atoms and molecules of which hold to each other tightly by chemical forces and do not require a gravitational pull for the purpose.

That means that particularly hot bodies like Mercury,

or particularly small bodies like the Moon, must be entirely solid and can have no oceans.

For an ocean to exist, a planet must be large enough and cool enough for the purpose, or warm enough. The requirements are rather stringent.

Thus, Mars is larger than Mercury and cooler, so that it can hold some volatiles—but not enough for an ocean. And Mars is sufficiently cool so that its volatiles are mostly in the frozen state. Venus, on the other hand, is even larger and has much more of the volatiles than Mars does—but it is so warm that all the volatiles are in the gaseous state. Under its thick atmosphere, Venus is a solid ball and has no oceans.

A planet as large as Venus or a trifle larger, one which is also considerably cooler, could retain ocean-sized quantities of volatiles and have much of it in the liquid form. But then, which volatile would form the ocean, or could the ocean be a mixture?

Suppose a planet is small enough to lose the hydrogen surplus, but is large enough to retain the volatiles. In that case, minus the hydrogen, there are chemical processes that tend to cause ammonia to become nitrogen (which remains gaseous) and water. There is also a tendency for methane to become carbon dioxide (which remains gaseous) and water. There is, finally, a tendency for hydrogen sulfide to become sulfur (which is solid and combines with other solids in the crust) and water.

Such a planet is therefore left with only *one* volatile in ocean-sized quantities and that is water. —And that is why the Earth is as it is.

Now what about the objects that form far from the distant star?

There the small atoms and molecules of helium, neon, and hydrogen are sluggish enough to be captured and, in their overwhelming presence, the mass of the forming body can increase rapidly. With increasing mass, the gravitational field grows more intense and the small atoms and molecules can be held even more efficiently.

The result is a giant planet made up very largely of hydrogen. Solid components, if any, make up an inconsiderable fraction at the center, and we have what used to be called a "gas giant." To be sure, it is now thought that Jupiter, although mostly hydrogen, compresses that gas into a red-hot liquid and that the giant planet is an enormous liquid sphere. It might be considered *all* ocean, but that is not the ocean in our sense of a partial liquid cover of a solid planet with dry land emerging here and there. No giant planet can have an ocean in our sense.

The far reaches of a planetary system need not contain only giant planets, however. Minor bodies are formed also out of leftover materials, and these can be as small as or smaller than any of the bodies of the inner planetary system.

The small bodies that are distant from the central star are cold, but even so don't have gravitational fields capable of retaining hydrogen, helium, or neon. Since the giant planets have swept up most of those substances, the smaller bodies cannot grow to the point where their gravitational field passes the critical point and a snowballing effect begins. Nevertheless, the small bodies of the outer planetary system can hang on to the volatiles, but the temperatures are so low that ammonia, water, and hydrogen sulfide, if present, will be there only in solid form. In the particularly far reaches even argon and methane will be frozen.

The result is that the small bodies of an outer planetary system are generally a mixture of ordinary solids (silicates and iron-nickel) and of "ices" (frozen volatiles). This is true, in our own Solar system, for instance, of the satellites of Jupiter, and of the comets.

It would seem, then, that the small bodies of the outer planetary system cannot have an ocean either—unless here, too, there are stringent conditions which, if met in just the right way . . .

The possibility arises in connection with methane, which boils at a temperature of $-161.5°$ C. Bodies of the nearer portions of the outer planetary system would be warm enough to keep it a gas; bodies of the outermost portions would keep it a solid. What about the region in between?

Suppose we have a body at just the right distance from the star to keep methane in the liquid state. If it is large enough to hold methane and not large enough to hold hydrogen, it may gather enough methane to possess a fairly thick atmosphere of that substance—with some of it liquid at the body's surface.

In fact, there's something more than that. Unlike the case of the other volatiles, the molecules of methane can, under certain conditions, combine into larger molecules which could be liquid even while methane itself is gaseous. These larger molecules would be "hydrocarbons," rather like lighter-fluid in nature.

As it happens, there is a body in our own Solar system which just possibly may qualify in this respect. It is Titan, the largest satellite of Saturn, and, in point of volume, the largest in the Solar system. Titan has a fairly thick atmosphere (the only satellite known to have a sizable one) containing methane.

Has Titan, then, a hydrocarbon ocean covering much of its surface? We can't say, but it is, at least, conceivable.

To summarize, then, for an astronomical body to have an ocean on its surface, it must fulfill very stringent conditions in terms of size, temperature, and gravitational intensity. so that only a small proportion of the planetary bodies in the Universe may be expected to have one.

On the other hand, any astronomical body that happens to be just about Earth's size and temperature is almost sure to have an ocean, and that ocean is bound to be water.

Furthermore, there is a chance that an astronomical body that is Earth-size and somewhat smaller, and that is much colder than Earth, could have the only other variety of ocean that is even conceivable—one of hydrocarbon.

The score for our own Solar system is one water-ocean (Earth, or I wouldn't be here to write this article, or you to read it) and just possibly one hydrocarbon-ocean (Titan).

9 TECHNOLOGY AND ENERGY

FOREWORD

We move on, now, to a pair of articles on physics.

Back in the fall of 1973, a Chicago engineering firm, alarmed by the spreading distrust of technology among the population, wished to undertake a series of pamphlets arguing the case in favor of technology. I agreed to do the first two pamphlets and they duly appeared and were reprinted in my collection Science Past—Science Future *(Doubleday, 1975) under the titles "Technology and the Rise of Man" and "Technology and the Rise of the United States."*

It was while I was writing these two articles that the Arab oil embargo hit the United States. Technology suddenly became a far more crucial matter to the average American and I was not surprised when the engineering firm eventually came to me for two more articles, one of which was to be primarily on energy.

That one, which appears below, appeared in pamphlet form under the rather interesting title "Technology: Whale Oil, Arab Oil and No Oil." It appeared at the end of 1975 and was the fifth pamphlet in the series.

The major feat of technology is just this—that it has made more energy available to human beings. Even a rather poverty-stricken American of today has at his disposal more energy of more kinds than any Alexander or Caesar of the past can possibly have had. And while energy does not, of itself, bring happiness, lack of energy is very likely, of itself, to bring misery.

The increase in energy has been accomplished in three different ways:

1. New sources of energy have been made use of, as when coal began to be used to supplement the burning of wood; and oil to supplement the burning of coal; and fissioning uranium to supplement the burning of oil.

2. Old sources of energy have been used in greater quantity, as when new oil fields are discovered and exploited, or a river is dammed to provide a new source of hydroelectric power.

3. Sources of energy, old or new, have been used more efficiently, so that a smaller percentage of the energy is wasted and a larger percentage put into useful work.

The first two methods of increasing the energy supply are noticeable and much noted; obvious, visible, and much talked of. The third method usually goes unsung, unheeded, unheralded.

So obvious is it that the total amount of energy being used per unit of time has been rising precipitously over the last two centuries that it obscures the fact that the energy is being used less and less wastefully.

Let us consider an example from an advance made two centuries ago.

Our present high technological civilization is usually traced back to the invention of the steam engine by James Watt. By means of this device, heat could be turned into useful work on a large scale. The power of expanding steam came to replace human and animal muscles in mines, in factories, in transportation, and thus began what is called the "Industrial Revolution."

But what was it James Watt invented? He certainly did not discover a new source of energy, since his steam engine would work perfectly well if the steam were produced by the use of burning wood—humanity's oldest fuel. Nor did he even invent a completely original way of handling energy, for his steam engine was *not* the first useful device of its kind.

Seven decades before Watt's invention, two engineers, Thomas Savery and Thomas Newcomen, were attempting to build steam engines. Savery constructed a device in which steam, once formed and confined till it reached high pressure, could be used to lift a column of water out of the depths of a mine and blow it out to ground level. Such a device could be used to pump water out of mines, but under the then state of the art, high-pressure steam was dangerous. The engine could explode and kill.

Newcomen made use of steam at ordinary air pressure only. A chamber was filled with heated steam and it was then cooled. The steam condensed and water was sucked into the vacuum thus produced. This process was repeated over and over and water was gradually sucked out of the mine.

By 1712, Newcomen had built an adequate steam engine and by 1725 he was selling them to mine owners. In 1778 more than seventy Newcomen engines were pumping away in the mines of Cornwall alone.

In 1764, the University of Glasgow asked James Watt to repair a Newcomen steam engine that had broken down. This Watt could do without trouble, but once he had the steam engine working he couldn't help observe how terribly inefficient it was. Less than 1 percent of the energy of the burning fuel, used to convert water into steam, ended up doing the work of pumping water. More than 99 percent of the energy simply flooded into the environment as unused heat.

Eventually, Watt perceived the chief source of inefficiency. In the Newcomen engine, there is a single chamber which fills with steam and grows hot. The chamber is then cooled to condense the steam and produce the vacuum. It then had to be filled with steam again, but any steam that entered condensed until such time as the chamber grew hot again. A great deal of steam was first necessary to heat up the chamber before additional steam could be used for work. Large quantities of steam were wasted at every cycle and immense quantities of fuel were required to undo the work of the cold water, and to heat and reheat the chamber.

Watt introduced a second chamber into which the steam could be led from the first chamber. The second chamber was kept cold constantly while the first was kept hot constantly. In this way, the two processes of heating and cooling were not forced to cancel each other. Furthermore, since there was no long pause at each cycle to heat the chamber, Watt's engine did its work much more quickly.

Watt devised other improvements. He allowed steam to enter alternately on either side of a piston, instead of on one side only, thus greatly accelerating the speed and ease of push-pull. He devised mechanical attachments that converted that back-and-forth movement of the piston into the rotary movement of a wheel, making his steam engine infinitely more versatile than Newcomen's had been.

By 1800, five hundred Watt engines were working in England and the number increased each year. Their ability to turn wheels meant that they could be set to running machinery of all sorts in England's textile factories. So successful was Watt's engine that the general public forgot the Newcomen engine had ever existed and Watt is generally considered the inventor of *the* steam engine.

Yet although the Watt steam engine conserved energy to a far greater extent than the Newcomen engine did, this fact was not easily visible. The mere fact of its greater economy meant that such huge numbers of Watt engines came into use that the *total* amount of energy consumed was vastly increased, and this was what was really noticeable.

The tale of the Watt steam engine is that of one industrial device proving more efficient and energy-saving than another. It is also possible to demonstrate that technology can use energy more economically than a *pre*-industrial process can when both are accomplishing (or attempting to accomplish) the same process.

What is so important to human society as communication? The development of speech—that is, communication through modulated sound in so subtle and versatile a manner as to make it possible to express abstractions—is a property of *Homo sapiens* only, as far as we know (though some speculate that dolphins have the ability, too). It is speech, and the methods for recording it, that make the difference between an animal and a human being.

As long as direct speech was the sole important means of human communication, two people could communicate directly only at distances over which the voice could carry. If a communication had to be sent from one person to another who was out of hearing range, this could be done only by having the speaker, or a hired messenger, move bodily across the distance in order to relay the message. Pheidippides had to run twenty-six miles to tell the Athenians the Persians had been defeated at Marathon, and died in the marketplace after delivering that message.

With the existence of writing, a message could be sent of indefinite length without fear of distortion in the process and with some assurance of secrecy, but that message still had to be carried the full distance by some human being if it was to be delivered.

Yet society could not do without the delivery of such messages. One might argue that a soldier or merchant far from home who wishes to be assured of the health of his loved ones must simply do without the assurance. Nevertheless, if any realm larger than that of a city-state is to exist, there must be messages from all parts reaching the central directing body, and other messages moving out from the central directing body to all parts—to say nothing of messages sent out to other realms.

Yet through all the great empires of human history from Egypt and Sumeria, through Rome and China right down to the nations of the century before our own, all messages had to be carried by a man, who might be walking, or sailing, or on horseback, galloping.

From the time the horse was first tamed on the steppes of Central Asia four thousand years ago, right down to the 1800s, the galloping horse was the fastest manner in which a message could be carried on land, the wind-driven ship the fastest manner in which a message could be carried across the sea. Couriers held the Persian Empire together in 500 B.C. by galloping in relays, and the pony express bound the coasts of the United States together in 1860 by galloping in relays. As late as 1814, it took so long for news to cross the Atlantic that the greatest battle of the War of 1812 was fought at New Orleans three weeks after peace had been signed in Ghent.

Even the replacement of the horse by the railroad and the sailing vessel by the steamship didn't speed things up much.

Was there any way of sending a message without physically carrying it? A mirror might reflect Sunlight in coded patterns, or a semaphore on a hilltop could manipulate the positions of its arms, but that requires daylight, is clumsy, and lacks privacy.

The use of an electric current along a wire was the answer. In 1844, Samuel F. B. Morse sent the first telegraphic message, "What hath God wrought?," from Baltimore to Washington. First the nation and then the world began to be wired for message conduction and reception. By 1866, cables were stretched across the Atlantic, and the United States and Great Britain could communicate in seconds independently of the motion of ships and men. In 1876, Alexander Graham Bell patented the telephone, which could transmit a replica of the human voice, rather than a mere code of dots and dashes.

Consider the saving in energy involved in a three-minute call between New York and Los Angeles as compared with the physical transmission of that same message by someone walking, riding, going by automobile, train, or airplane. There is no way of duplicating what the delicate surge of the electric current does in any other fashion without enormously increasing the energy expenditure.

In one way, in fact, the energy saving is beyond calculation; it is actually infinite. The message by phone races from one point on Earth to any other point in less than a tenth of a second. If the message must be physically carried, then there exists no way, no way at all, whatever the expenditure of energy, of delivering it as quickly.

As it happens, I am cheating a bit. The energy cost of a telephone call from New York to Los Angeles can't be calculated by taking into account the electric current along the wires only. What about the wires themselves, and the telephone poles, and the underground cable, and the telephone exchanges, and the telephone employees?

The expense is enormous.

To walk from New York to Los Angeles in order to deliver your message would be much cheaper than what it would cost you if you were charged for the construction of the entire telephone system.

However, there was never any need to build a telephone system in order to transmit one message. Once people discovered that communication by wire was available, they also discovered they had a great deal to say and a great deal they wanted to hear.

If messages were few, over long distances, before the days of electric communication, that was not because people had little to say, but because the time, effort, and uncertainty of delivery discouraged them into silence. Once time, effort, and uncertainty were all reduced to near zero, the tide of talk rose.

To build a telephone system that will suffice to carry a million calls per second is not a million times as expensive as building one that will carry one call per second, nor does it take a million times as much energy. In fact, cost and energy rise so slowly with increasing numbers of messages that at a certain point the use of electric communication represents an enormous saving in energy over what had existed previously, even if all possible expenditures in setting up such a communication system are counted in.

Imagine, for instance, that all telephone conversations in the world suddenly cease, and that all messages that would have gone out over the telephones in the course of the next twenty-four hours must be delivered personally, either in speech or in writing. The energy expenditure involved in such personal delivery would be so great that not only could the messages not be delivered in time, they probably could not be delivered at all.

The only reason, in short, that the world can retain its intricate system of intercommunication at its present level is that technology has devised a way of doing it at a low enough cost of energy per message to make it possible.

Again, I may be cheating a bit. Is it fair to compare the level of electric communication with a personal-delivery system that attempts to remain at the same level? Wouldn't it be true that if electric communication were abolished, the world would once again reduce its messages to a number that could be carried on by personal delivery under a reasonable set of conditions?

It might be argued, in fact, that electric communication has raised the level of human talk to an entirely unnecessary height; that most of the talk is unnecessary babble, wasteful babble, senseless babble, and should indeed be choked off. It might be argued still further that electric communication has accomplished so little that it has in no way replaced the personal delivery of messages by speech or writing; that people travel on business now more than ever, and that the volume of mail has risen steadily through all the years since electric communication came to pass.

The answer to all this is that, even allowing for unnecessary babble (the existence of which is impossible to deny by anyone who has ever had a teen-age daughter and a telephone), technology has made it possible to support a much more intricate and elaborate social and economic structure than would have been possible otherwise. The more intricate and elaborate structure may be troublesome to maintain and may produce enormous problems, but it has also made it possible for the world population, and, particularly, for the American population, to enjoy a greater material prosperity and security than anything ever before seen in history. (And even the communication of babble adds to the happiness of the babbler.)

In turn, the rising tide of population and prosperity has produced more people with more things to say and more desire to read—more people with money to spend and therefore more advertising to induce them to spend, to say nothing of more worthy causes soliciting funds and politicians soliciting votes. Naturally, the use of mail goes up even as that of the telephone does.

In short, if we imagine a cessation of all electric communication, we don't merely wipe out idle talk, we also make it impossible to maintain society at its present complex level. All at once it must take on a much simpler complexion.

Are there those who say this is good? Certainly. There are many people who (if they are to be believed) long for the simple life and would just love to live on a farm and grow their own food and do away with machinery and telephones and stare at the Sunset. No doubt they can do it, if there aren't too many of them, and if they have the comfortable knowledge that they can always find some particular telephone or other technological device, or a hospital, in case of emergency.

If *everyone* abandoned electric communication and not just the occasional simple-life elitist, then, for one thing, the Earth simply could not support the present population, which is at least four times as high as can be supported by any form of social and economic structure markedly less complex than presently exists. (There were less than 1 billion people on Earth when Watt devised his steam engine, and that billion lived shorter and more miserable lives—on the average—than today's 4 billion do.)

The fact does remain, though, that every technological advance that makes something less energy-expensive to do makes that something more available to the general public, with the result that its use is enormously increased and the total energy expenditure rises precipitously. If we cannot reverse gears and, without disaster, abandon the technological advance once it has become an integral part of the social and economic structure, what can we do?

There are three things that might be done:

1. The population increase must level off and then be brought down to some bearable level. No matter how bare an existence each additional person can be forced to maintain, he or she will need a certain amount of food and shelter if life is to be possible at all, and this represents an energy expenditure that, in the long run, cannot be supported.

At the present rate of population increase, there will be 7 billion people on Earth in 2000 and 50 billion in 2100. Somewhere between now and 2100 (and probably closer to now than to 2100) it will become impossible to find energy enough to support the world's population at even the lowest level, since we will also have to be using energy to maintain a social and economic system intricate enough to get the food and shelter to all these people.

This is not a purely technological problem. Technology, by making possible contraceptive devices (including, particularly, oral contraceptive pills), has done much to bring the birthrate down to replacement levels in the United States and in other technologically advanced countries, but that is not enough. There are political, economic, and social factors that make the population problem enormously intractable, and one which may not be solved fast enough to prevent catastrophe.

Technology has an important negative influence, however. If it falters, because of attacks from the simple-lifers, or for any other reason, so that the social and economic system loses what ability it has to make use of and to distribute the resources of the world, the catastrophe will come the sooner.

2. The philosophy of energy conservation must become part of the human way of thinking. Whether by laws or by the pressure of public opinion, needless energy waste, either by consumers or by producers, must be stopped.

3. Technological advance must continue in the direction of saving energy, of reducing the energy expenditure per task accomplished, even as it proceeds to uncover large supplies of old energy sources and of altogether new energy sources.

And if one sometimes despairs of anything ever being

done about population and waste, it must be admitted that technology has been constantly making energy do what it could not do before, and making it do more per fixed amount than it could do before.

Forms of electromagnetic radiation other than light were discovered. In 1888 Heinrich Hertz learned how to produce and detect radio waves, and by 1901 Guglielmo Marconi was using them to send messages across the Atlantic Ocean. Radio made it possible to send messages without wires, and to that extent the communications network was, at least potentially, greatly simplified. Radio also extended communication, for the first time, to ships at sea and planes in the air.

Wires grew less important in another way. By the 1890s it was recognized that the electric current involved particles smaller than the atom, and in 1895 J. J. Thomson discovered the electron. Under the proper conditions, a stream of electrons could be sent through a vacuum and that stream could be deflected there, started, stopped, diminished, intensified.

All that could be done with ordinary mechanical switches that opened and closed electric circuits could be done by manipulating electrons in a vacuum—and, in the second case, much more quickly and much more delicately since electrons were so incredibly light. Such an electronic switch was developed by men such as John A. Fleming and Lee De Forest, so that by 1906 what is commonly called the "radio tube" was in existence.

These radio tubes made possible a host of "electronic" instruments. At the cost of a tiny expenditure of energy spent in manipulating the infra-light electrons, those instruments did work that would have cost much more energy in non-electronic fashion—if it could have been done at all. Radio tubes made possible not only radio but also television, record players, electronic computers, and so on. The world, which had grown electric after 1844, grew electronic after 1906.

The radio tube is by no means incapable of improvement, however. Tubes must be fairly large since they must enclose a quantity of vacuum through which the electrons travel, and they therefore contribute the major portion of the bulk of electronic instruments. They must be heated before the electrons can be driven out of the wires and into the vacuum and that takes time and energy. They

81

will eventually develop leaks, so that they will have to be replaced.

As it happens, electrons can drift their way through certain pure solids or mixtures of solids and there they can be controlled precisely as in a vacuum. In 1948 W. B. Shocklay, W. H. Brattain, and John Bardeen developed the solid-state equivalent of a radio tube and called it a "transsistor."

A transistor, being solid and possessing no vacuum, does not leak and is sturdy enough to last indefinitely. Not enclosing a vacuum, it can be very small, and since it does not have to be heated, it can begin work at once and consumes very little energy. As electronic devices become transistorized, they become smaller and less energy-consuming. Radios and fairly complicated computers can now be made small enough to fit into a pocket.

More remarkable still, satellites measuring only a few feet across can carry an enormous load of sophisticated instruments that are capable of measuring the most delicate properties of space, or of taking detailed photographs of planets and then sending back the information across hundreds of millions of miles of space.

In the days before transistors, satellites would have had to be enormous to carry instruments capable of accomplishing all this, and they therefore couldn't have been launched at all since no one would have cared to support the energy expenditure—and by "the days before transistors" we mean only thirty years ago.

Even in this transistorized and miniaturized age, our greatest economies but encourage the still further growth of total energy expenditure. The vast growth of radio has not prevented a continuing growth of the telephone network, for instance.

It must also be remembered that the continuing increase in total energy expenditure also results, in part, from the fact that Earth's population since the coming of the transistor has nearly doubled, and that technologization is spreading over the hitherto non-industrial areas of the globe.

What of the future? Can technology continue to save on energy and reach a point, finally, where even the total expenditure is reduced? Can technology point the way to a less energy-intensive world altogether?

Consider—

The transistor has made satellites possible, and the communications satellite has increased the ease and decreased the energy with which it is possible now to communicate from continent to continent.

Since 1960 we also have laser beams in which visible light waves can be organized as tightly and as uniformly as radio waves can be. This means that laser beams can be used for communication eventually, and because light waves are a millionth as short as radio waves, visible light has room for a million times as many communication channels as radio waves do.

With communication satellites and laser beams working at full efficiency and with advanced computers organizing the whole, it may be that the world can begin to transmit messages with an efficiency which will make ordinary radio (to say nothing of the telegraph) look like a remnant of the days of the pony express.

With communications satellites and laser beams working nels available, every man could have his portable phone, and dial any number on Earth. The printed word could be transmitted easily and widely, so that facsimile mail could be sent from point to point in a fraction of a second. Facsimile newspapers, magazines, and books could be readily available at a press of a button. Perhaps, eventually, a single world computer will hold in its vitals the library of mankind, any part of which will be available to any man at any time.

It is this enormous advance in ease of message transmission which may finally represent the kind of advance necessary to reduce total energy expenditure.

It would no longer be necessary to transfer mass when an image could be transferred instead. Documents would not have to be carried by a person who would, in turn, have to be carried by a plane. The documents themselves could be converted into radio signals and reassembled into facsimile documents at the other end, with an enormous saving of energy.

For that matter, the image of an individual can itself be transferred, rather than the individual himself. Conferences can be held even though those making it up are separated by global distances. The conferees may each be seated at home, with those homes scattered over six continents and three dozen nations, and yet with all their images together.

No one would need to be at any one particular spot to

control affairs. Businessmen would not need to congregate in offices, nor workmen in factories. With automation, telemetering, and television monitoring, all work could be done from home if that be desired.

To be sure, mass would still have to be transported where that mass represents food, raw materials, finished products, but so much mass would not need to be transported just for the information it contains that the savings would be enormous.

People would still travel, but only for pleasure and not out of business necessity—and the pleasure would be the greater since the means of transportation would not be crowded by those whose purpose was other than pleasure.

Cities themselves would spread out and disappear. There need be no congregation for ease of communication or for business reasons. There won't even be any need to congregate for cultural reasons, since all books, plays, concerts, and everything else of the sort will be equally available anywhere. Every place on Earth will be "where it's at."

And this means that *provided*—

1. population is stabilized at a reasonable level and that

2. people learn to avoid waste and useless destruction of resources while technology is developing endless new sources,

then it will become possible for human beings actually to have a form of the simple life, with space, privacy, and dignity, for an indefinite time to come—all mediated, as it must be, through technological advance and the wise use of energy in the most economical fashion.

Nothing like this could *ever* be brought about by any suicidal turning away from or rebellion against technology. With technology defeated or aborted, there can only be rapid catastrophe and the permanent debasement of what will survive of the human spirit.

AFTERWORD

The remaining article, which I entitled "Technology and Communication," was bought and paid for and was supposed to be out in the summer of 1976. Despite several inquiries, however, I have seen nothing of this one. For that reason I can't include it here. If it finally does appear, however, it will go into some future collection.

10 THE GLORIOUS SUN

FOREWORD

In the summer of 1974, I was asked by American Way, *the inflight magazine of American Airlines, Inc, to do a monthly column for them. It was to appear under the general heading of "Change" and I was, each month, to speculate on one aspect or another of the future. I gladly agreed because I thought it would be fun—and it has indeed proved to be so. I haven't missed an issue since that of November 1974.*

Just a matter of days before I agreed to do the column, however, I had agreed to do an article on solar energy for East/West Network, which publishes a number of competing in-flight magazines. This struck me as a conflict of interest but, having agreed, I couldn't back down. I consulted John Minahan, who was editing American Way *then, and he assured me it was perfectly all right in the most cheerful manner possible.*

We live on the energy of the glorious Sun; all of life does. The green plants make use of the energy of Sunlight to convert carbon dioxide, water, and minerals into carbohydrate, fat, and protein.

The animals live on the high-energy compounds of plants, or on other animals that have eaten plants. All animal life, ourselves included, feeds ultimately on the green plants that have used the energy of Sunlight to create the food supply.

Man's technology is based on the energy of the Sun, too. The Sun's heat warms the air and sea unevenly, creating wind and ocean currents. The Sun's heat evaporates the ocean, lifting cubic miles of water into the air in vapor form. There they eventually condense and fall as rain; some of it on the continents, where they collect in ponds and lakes, and run back to the sea in streams and rivers. And the winds and running water have been moving ships and turning wheels since ancient times.

The great man-made energy source, fire, depends on the burning of fuel in air. Where the fuel is wood, it represents the burning of compounds formed by plants through the use of the energy of Sunlight; where it is animal fat, it is the compounds formed by animals at the expense of plants; where it is coal or oil, it is material that was formed by plants or animals hundreds of millions of years ago out of the energy of that ancient Sunlight.

Some man-used energy is not of Solar origin—the internal heat of the Earth shows up in hot springs, the Earth's rotation produces the moving tides, and atomic nuclei can undergo fission or fusion to produce energy.

These non-Solar energy sources contribute, so far, very little to the total energy needs of mankind. The major source at this moment (and for two centuries past) has been the coal and oil obtained from within the Earth's crust.

Coal is hard to get, however, and hard to transport, and digging it out of the ground does damage to the environment. Oil is in limited supply and the day of its disappearance is not too many decades into the future. Both coal and oil, in burning, pollute the air badly.

Even if coal and oil were properly purified and burned with complete efficiency, so that no ordinary pollution was produced, they would still yield waste heat that would slowly warm the Earth and alter its climate. They would also produce carbon dioxide that would keep that heat from escaping into space and would accelerate the warming trend.

If we switch to nuclear fission, there is the great danger of radiation pollution. If we switch to nuclear fusion, where the danger of radiation pollution is much less, we must face the fact that the engineering problems involved in fusion are as yet unsolved and may take decades to solve.

We can turn once again to the Sun. Despite all the Solar energy that goes into producing wind, water currents, and green plants, over 90 percent of it goes into merely heating the Earth. That heat is useful of course, for it keeps the temperature of the Earth warm enough to make life possible. Still, if that waste Sunlight were used for man's purposes, it would still end up as heat (which is indestructible) and Earth would remain as warm as before.

Each *day* the amount of Sunlight that strikes Earth without being used in any way but to warm our planet represents as much energy as mankind is now using in three *years*. What's more, Solar energy is completely nonpolluting. It doesn't even introduce heat pollution, since the heat is there in the same amount whether we make use of the energy or not.

What keeps us, then, from making use of this Solar energy? Three things—

1. Solar energy is very dilute. There's a lot of it, but it is spread thinly over a large area. Collecting it, and concentrating it to the point where it is useful for man's technology, is tricky.

2. Solar energy varies in amount. It is low in the morning and evening and nonexistent at night. Clouds, mist, and fog cut the amount even when it is at its maximum. In many places where man's industry is most concentrating it to the point where it is useful for man's cularly variable.

3. People have been too lazy to work out the engineering problems involved in the direct use of Solar energy, as long as the easier techniques of burning coal and oil were available—and too unimaginative to see the necessities and possibilities far enough in advance.

Solar energy can be used on a very small scale. Sunlight beats down on the roof of each building. If the roof is covered with black panels to absorb the heat, and if this heat warms water in shallow tanks just beneath the panels, and if this water is circulated, the whole building can be heated in cold weather. For that matter, heat energy can be used to run an air-conditioning device that will cool it in warm weather.

The Sun may not be dependable enough to keep such a "Solar house" going during cloudy spells or when the weather is extremely hot or cold, but more conventional backup energy sources could be used—very small quantities of those in the long run.

Why, then, is this not done?

In the first place, it is. Sun-powered homes are built here and there, particularly in Japan, but, on the whole, only occasionally and only experimentally. The initial cost is high and the construction industry is reluctant to invest the money as long as the public, not clearly understanding the saving in the long run or lacking the short-run capital, is not eager to buy.

Other small-scale uses involve Solar stills in which Sunlight is used to evaporate sea water, so that fresh water can be condensed and collected, and Solar furnaces in which Sunlight is reflected by an array of mirrors and is focused at some point where the temperature then reaches near to that of the Sun's surface.

The energy of Sunlight can also be used to produce electricity, a more flexible and delicately useful form of energy than mere heat is.

Certain wafers of metals, of carefully adjusted composition, will give rise to a small electric current for as long as they are exposed to Sunlight. Such "Solar cells" have been used to power artificial satellites with great success.

Imagine banks of Solar cells lined up on rooftops or on other blank surfaces exposed to Sunlight. Electricity might be produced in steady quantities that would run appliances. It could be stored in batteries and used to light buildings at night.

To be sure, Solar cells are quite expensive and are rather fragile. At the present moment, Solar electricity would be some five hundred times as expensive as electricity produced in more conventional ways.

But then, Solar cells have till now been produced in small quantities for specialized purposes. If efforts are made to produce more rugged cells in mass-produced fashion, the price should drop drastically.

We could then imagine huge power plants based on a vast array of Solar cells covering large areas of those sections of the Earth where Sunlight is most nearly continuous. These sections happen to be desert areas where there is little life, and where Sunlight merely heats bare sand and rock uselessly.

About 7.7 million square miles of the Earth's surface is in the form of hot Sun-drenched desert. The Sahara Desert alone is nearly as large as the United States. Solar cells, working at only 10 percent efficiency, would require 30,000 square miles of Sunlight (only 1/250 of the world's desert area) to supply the present energy needs of the world. Here in our country, we have large sections of the Southwest we could use as a source of Solar energy.

Naturally, this would demand a large initial investment, and it might be the oil sheiks who could lead the way.

At the present time they are gathering the wealth of the world into their hands in return for the oil which they own, and are in somewhat of a quandary as to what to do with that wealth.

Surely, the oil-producing nations of the Middle East are well aware that their oil is a dwindling resource and that, as it happens, their lands also contain generous sections of the Earth's Sun-soaked deserts. If they are farsighted enough, they will finance the research and engineering that would turn their lands into centers of Solar energy. They would, in this way, conserve their economic strength even while helping the rest of the world, which could use the Middle Eastern experience to build power stations in desert areas elsewhere.

The power stations based on Earth's deserts are perhaps not the ultimate. Earth's atmosphere reflects more than half the energy of Sunlight back into space, before it reaches Earth's surface. The atmosphere absorbs some of what is left. Then, too, deserts have their sandstorms and could be subject to damaging earthquakes. The mere fact that the power stations would be on Earth's surface would mean that they would interfere with life forms, including man; and vice versa.

There are suggestions, then, that the power-collecting devices may someday be taken off Earth altogether and be assembled into various orbiting Solar-power satellites. These could absorb Sunlight without interference by nighttime and without loss by Earth's atmosphere. The energy they absorbed could be beamed to Earth in the form of microwaves (like those that are used in radar) and on

89

Earth those microwaves could be picked up by huge antennas.

Thirty-six years ago, I wrote a story* picturing such power stations circling the Sun in the neighborhood of Mercury's orbit, where the Solar energy is nearly ten times as concentrated as it is here, near Earth. (The power-station satellites were run by robots in my story.)

The concept was pure science fiction then, and is still science fiction now—but in the intervening third of a century it has come considerably closer to practicality. When I first wrote the story, no one except science fiction writers dreamed of satellites and space stations at all, and scientists were just beginning to learn how to handle microwaves.

Given another third of a century, and who knows—

What we need is the ability of scientists and engineers to overcome the practical problems in the way, the resolution of political leaders in backing them, the ability of the people generally to understand the potentialities of the direct use of Solar energy and their willingness to see their tax moneys used for the purpose, and, most of all, the continued stability of the world's social, economic, and technological system.

We need vision; we need courage—and we need some luck, too.

AFTERWORD

The above article appeared in Mainliner, *the in-flight magazine for United Airlines, at just about the time my second* American Way *column appeared. As a result, several of my more airplane-prone friends said to me, captiously, "Good God, Isaac, must I read you on every airline I take?"*

---

* "Reason," *Astounding Science Fiction*, April 1941.

11 ASTRONOMY

FOREWORD

Writing this aricle was a touch frustrating. I had been asked to do so by Family Creative Workshop, *a young people's encyclopedia, and here I am urging everyone to look at the sky—and I don't. Living in New York, there's no point. The city lights and the city smog wipe out almost everything. (Janet, who is a more determined viewer than I am, uses a pair of good binoculars to study Jupiter every chance she gets. Each time she sees the four satellites it is a triumph.)*

When we think of astronomers nowadays, we think of huge telescopes and cameras; of advanced instruments. We think of satellites in the sky, recording data from the ends of the Universe. You and I have none of this. All we have is our eyes; and all we can do is look.

Yet, until three centuries ago, that was all anyone could do. Over a period of thousands of years, all the astronomy mankind could work out was based on what he could see with his unaided eye.

Early man had an advantage, of course. In his day, the air was not yet filled with the dust and smoke of an industrial civilization. Bright city lights shining all through the night did not yet drown out the stars. It is hard for us today to find a really dark place with really clear air, where we can look up and study the night sky. (We can get an idea of what we might see if we could do this, however, whenever we visit a planetarium.)

Still, all is not lost. If you live outside a city, put out the lights in your house and go out into the back yard on a clear Moonless night. You will see the stars. Even in the city, if you go to the roof of an apartment house late at night and find a place where the night glare of the street-lights is least, you will see the stars.

You may see only the brighter stars, but even so you will surely find objects of interest, and in this article I will point out a few of them.

If you live in the United States, then, on any night of the year, you will see, somewhere in the northern part of the sky, the seven bright stars that make up the Big Dipper. This is the best-known star group in the sky. Everyone knows the Big Dipper.

For one thing, it *looks* like a dipper. There are four stars making up the bowl of the Dipper, arranged in a slightly uneven rectangle. The other three stars, in a bent line, make up the handle of the Dipper.

If you were to watch the Big Dipper all night long, you would see that it seems to move in a large circle around some point in the sky. Actually, the whole sky seems to move because the Earth is rotating. This is most noticeable in the case of the Big Dipper because its distinctive shape makes it easy to watch.

Someimes it is seen in the position you would expect a dipper to be. Sometimes it is standing tipped on its bowl, sometimes on its handle, sometimes it is upside down, *but it is always above the horizon.*

This means the Big Dipper is always visible on every cloudless night and is always in the northern region of the sky. This may have been the first great astronomical discovery of mankind. The ancient Phoenicians, about 1000 B.C., may have made use of this fact to guide themselves across the open sea. As long as they could see the Big Dipper, they knew which direction was north, and could work out all the other directions.

The Phoenicians were the first people who dared strike boldly across the Mediterranean Sea and to explore it from end to end. They were the first to make their way out of the Strait of Gibraltar into the Atlantic Ocean—guided by the Big Dipper.

The seven stars of the Big Dipper have names that are Arabic in origin. This is because the Arabs were the great astronomers of the early Middle Ages. When the Western Europeans began to study astronomy again after A.D. 1000, they did it by translating Arabic books and making use of the names the Arabs gave the stars.

The name of the middle star of the handle of the Dipper, for instance, is Mizar (MY-zahr), which is an Arabic word meaning "veil." It is called this because its light veils a dimmer star very close by called Alcor (al-KAWR), from Arabic words meaning "the weak one."

Mizar and Alcor make up a double star, and it is very difficult to see the dimmer one. To the Arabs, it was a test of good eyesight to see Alcor. It still is today except that the glare of city lights and the dustiness of the air make the test unfair. I have never seen Alcor even though, with my glasses on, my eyesight is perfect.

Now concentrate on the two stars of the bowl of the Dipper that are on the side opposite the handle. These have Arabic names, too, but never mind those. They are commonly known as "the Pointers."

Why? Well, draw an imaginary line through the two Pointers and follow it past the open top of the Big Dipper. Continue that line about seven times the distance between the two Pointers and you will come to another bright star. This is the North Star.

The Big Dipper, as it turns in the sky, turns about an imaginary point called the North Celestial Pole. This point in the sky is directly over Earth's North Pole. Earth's North Pole stands still while Earth rotates. Earth's rotation west to east makes it appear that the sky is turning east to west, and the North Celestial Pole, like Earth's North Pole, stands still.

The North Star is very near the North Celestial Pole and makes only a tiny circle about it. The North Star seems to be always in the same place therefore and when you look at it you are looking due north.

If you continue the line of the Pointers past the North Star, you will come across five bright stars a little to one side of the line. This makes up the star group (or "constellation") known as Cassiopeia (KASS-ee-oh-PEE-uh).

Cassiopeia is the name of a queen in Greek mythology. The Greeks tended to imagine the stars forming groups that looked like human beings, animals, and other objects. These pictures they imagined were hightly fanciful and it doesn't really pay to try to make them out. The Greeks pictured Cassiopeia, for instance, as a woman sitting in a chair. For ourselves it is more useful to see it as five stars in the shape of the letter W.

Cassiopeia is always on the opposite side of the North Star from the Big Dipper, and both turn and turn and turn without ever setting.

Imagine a line extending from Cassiopeia down to the Earth, going through the center of the Earth and out to the sky on the other side of our planet. Near where that line would reach the other side of the sky is a bright star called Alpha Centauri (AL-fuh-sen-TAW-ree).

Alpha Centauri is the third-brightest star in the sky. It is also the closest star in the sky, only 25 trillion miles away. (That doesn't sound close, but all the other stars are much farther away still.)

Since, as seen from the United States, Cassiopeia is always above the horizon, Alpha Centauri, on the opposite side of the sky, is always below the horizon. The ancient Greeks and the medieval Arabs never saw Alpha Centauri, and we can't either unless we travel south to the tropics or, for a really good view, to Argentina or South Africa.

Imagine, though, that we were living on a planet circling Alpha Centauri. Most of the stars are so distant that a shift of our eyes from here to Alpha Centauri is too small a change to make any important difference. The stars would still make the same patterns in the sky—with one important exception. Cassiopeia would contain not five stars in a zigzag line, but *six*, and the sixth star would be our Sun.

If you were far out in the country, with no lights anywhere around, and if the sky were cloudless, clear, and Moonless, you would see a dim shining band passing through Cassiopeia. You would be able to follow that band toward the horizon in either direction. This is the Milky Way.

The ancients never knew what the Milky Way was, but we do. The telescope has shown that it is composed of myriads of very faint, very distant stars. The whole system of stars we see is shaped like a huge lens. When we look through the long axis of the lens, all the innumerable stars we see for hundreds and thousands of trillions of miles melt together to form a luminous fog.

The Big Dipper and Cassiopeia are examples of star groups that never set and are always in the sky. Farther from the North Star, the star groups make such big circles that they move beyond the horizon. They rise and set. They are only above the horizon part of the time.

This means they can't always be seen. The Sun blots out the stars in its half of the sky and the Sun changes its position slowly against the stars, making a complete circuit of the sky in one year. This means that those star groups which rise and set will be above the horizon in the daytime during half of the year and then they can't be seen. They are above the horizon during the nighttime the other half of the year, however, and then they can be seen.

The nights are longer in the wintertime and the air tends to be clearer on cold nights, so that it is much better to view the sky late on a winter night (if you are well bundled up). As it happens, the star groups that are visible in the night sky when it is winter in the Northern Hemisphere are particularly beautiful, and I will describe some of them.

The most beautiful of all the star groups is the constellation Orion (oh-RY-on). Orion is the name of a giant hunter in Greek mythology and it is one constellation in which we can almost see the pictures as the Greeks imagined it. In its center are three bright stars in a closely spaced straight line. This is "the belt of Orion." Above the belt are two stars representing the shoulders of the giant hunter; below the belt are two stars representing his feet. (There are also somewhat dimmer stars representing a club in one hand, a shield in the other.)

The brighter of the two stars above the belt is a first-magnitude star (which is what the twenty brightest stars in the sky are called) and it is named Betelgeuse (BEE-tul-jooz). You may be able to make out the fact that it is reddish in color.

Betelgeuse is a "red giant." It is a cool star with a surface that is not very hot or bright and it would be dim indeed if it were merely the size of our Sun. It is, however, a huge star that pulsates, growing larger and smaller like a beating heart, and is between 300 and 400 times as wide as our Sun. At its largest it is 350 million miles across. If

Betelgeuse were put in the place of our Sun, it would stretch out to include the space through which Earth passes.

The brighter of the two stars below the belt is another first-magnitude star, named Rigel (RY-jel). It is smaller than Betelgeuse, but far brighter. It is about 16,000 times as bright as our Sun would be if both were seen from the same distance.

If Rigel were placed where the Sun is now, Earth's oceans would quickly boil away and its crust would grow red-hot. We would have to go far out to the planet Uranus to find bearable temperatures.

From one end of Orion's belt are three dimmer stars set at right angles to the belt. These form the "sword of Orion." The middle star of the sword is hazy, for you are looking not only at a star but at a patch of dust and gas called a "nebula" (NEB-yoo-luh), from a Latin word meaning "cloud."

Through a telescope this tiny patch becomes the "Great Orion Nebula," a vast foggy cloud containing enough dust and gas to make ten thousand stars the size of the Sun. The Great Orion Nebula glows because it is heated up by the light of nearby stars.

If you concentrate on the stars marking the two feet of Orion and look in the direction opposite to Rigel, you will come across another star, which shines like a blue-white diamond in the sky. This is Sirius (SIH-ree-us), the brightest of all the stars in the sky.

Partly it is bright because it *is* bright. twenty-six times as luminous as our Sun. Partly it is bright because it is close. Of all the stars we can see in the skies of the United States, Sirius is the closest, being only 50 trillion miles away. (It takes a beam of light one year to travel nearly 6 trillion miles, so it takes light from Sirius 8½ years to reach us. For that reason we say that Sirius is 8½ "light-years" away.)

The only bright star closer than Sirius is Alpha Centauri, which is 4⅓ light-years away, but it is not visible from the United States. There are three other stars known to be closer than Sirius. but these three are very tiny and dim and can be seen only with a telescope.

Sirius is called the "Dog Star" because it is part of a constellation that the Greeks visualized as a dog. The Greeks thought that Sirius was so bright it must give off heat. In the summer when Sirius is invisible because it is above the horizon during daytime, the Greeks thought that its heat, added to that of the Sun, made the temperature very high. That is why we still talk of the "dog days" of midsummer.

Of course, this is just a legend. Sirius does not give off any perceptible heat.

If you look on the side of Orion that is opposite to Sirius, you will see another bright star, Aldebaran (al-DEB uh-ran), part of a V-shaped group of stars that marks the constellation Taurus (TAW-rus). The Greeks pictured this constellation as a bull, with Aldebaran marking one eye.

Now look beyond Aldebaran on the side opposite to Orion. There you will see a small group of dim stars. There are six of these (some people can see seven) all clumped together. These are the Pleiades (PLEE-uh-deez), which are the most remarkable cluster of stars visible to the unaided eye. With a telescope, you can see hundreds of stars in the Pleiades; those you see with the eye alone are just the six brightest.

Taurus is one of the constellations of the Zodiac (ZOH-dee-ak), which is the name given to a band of star groups that completely circle the sky. The Sun, in its apparent motion against the starry background of the sky, passes through the entire Zodiac, remaining in each constellation one month.

The Moon and the various planets are also to be found somewhere or other in the constellation of the Zodiac. The planets are small bodies that shine only by reflected light from the Sun, but are so close to us that they shine more brightly than the stars do.

The position of the planets is always shifting among the constellations of the Zodiac. To find them, you have to locate them first on a chart. However, if you see a very bright star in the western sky soon after Sunset, when no other stars are visible, it is surely the planet Venus. Venus is brighter than any star and can be ten times as bright as even Sirius. A very bright star that is high in the sky late at night is very likely the planet Jupiter. If it appears reddish in color, it is Mars.

Not all the constellations of the Zodiac are as noticeable as Taurus. Another that is present in the winter sky and is noticeable is Gemini (JEM-ih-nee), also called the

Twins. If you draw an imaginary line through Orion from Rigel up through Betelgeuse and beyond, you will come to two bright stars close together. These are Castor (KAS-tor) and Pollux (POL-uks), the names taken from a pair of famous twins in Greek mythology. That is what gives the name to the constellation.

Moving away from the Zodiac, look in that part of the sky between Taurus and Cassiopeia. The stars in that area make up the constellation Perseus (PUR-syoos). It contains two bright stars and of these the one closer to Taurus is Algol (AL-gol).

Algol is a most unusual star. Every three days, Algol loses about two-thirds of its brightness over a period of five hours, then regains it in the next five hours. It is a "variable star," the most noticeable one in the sky. We now know that it behaves in this way because there is a dim star circling Algol and every three days it gets in front of Algol and eclipses it.

To the ancient Greeks, the behavior was mysterious, however. Perseus, in the Greek myths, killed the demon Medusa, whose face was so horrible that the sight of it turned men to stone. The Greeks pictured Algol as representing the head of Medusa, carried by Perseus. The very name Algol is an Arabic word meaning "the ghoul." For all these reasons, Algol is called the "Demon Star" but actually there is nothing demonic about it.

Before concluding this section, let me direct your eyes to one more object. Draw a line from the North Star to the middle star of Cassiopeia and keep going. The group of stars you will see just beyond Cassiopeia make up the constellation Andromeda (an-DROM-uh-duh).

In the part of Andromeda nearest Cassiopeia is a curved line of five medium-bright stars. Near the middle one of these stars is a tiny hazy patch which you might just barely make out on a very dark and very clear Moonless night. This is the Andromeda Nebula.

If you can make it out, take a good look, for it is the most remarkable object you can see with your unaided eye. It is not really a nebula, not a cloud of dust and gas; it is, instead, a collection of *two hundred billion stars.* It looks like a small patch of fog only because it is so far away.

It is 2,300,000 light-years away—the very farthest object you can see with the unaided eye. It is actually the

Andromeda galaxy; as large an object as our entire Milky Way.

AFTERWORD

The frustration aroused by writing the above article helped push me toward buying a telescope (largely on impulse). It came about December 1975 and we've looked at a few things with it, but New York remains an impossible base for viewing and the telescope is too bulky to take along with us on trips out of town. We're planning therefore to get a more compact item that can accompany us without undue effort on our part.

12 THE LARGE SATELLITES OF JUPITER

FOREWORD

After Pioneer 10 passed Jupiter in December 1973, Jupiter became big news, and the findings were indeed sensational—to astronomers. To the general public the findings may have fallen short inasmuch as there were no grisly mysteries that would threaten us with an invasion of extraterrestrial monsters.

The Baltimore Sun, *caught up in the excitement, asked me to do a piece on the findings of Pioneer 10 in connection with Jupiter's satellites. I wrote the essay here included and gave it the very sober title I am using here. The editor, however, couldn't resist, and substituted a much more menacing title, to wit: "What's Hiding on the Moons of Jupiter?"*

Well, folks, in all likelihood, just ice and craters and things like that.

Pioneer 10 passed near Jupiter in December 1973 and, in doing so, did more than merely study that giant planet.

It also took mankind's first close look at the four large satellites that circle Jupiter.

As it happens, an unmanned look, by instrument, may be all we'll ever get of those satellites, for it turns out that Jupiter's magnetic field is far more powerful than had been thought. It pushes out from Jupiter's equatorial zone in a kind of doughnut shape that is four million miles across. Within it are enormous quantities of high-energy particles, and also within it are the four large satellites.

If men and women are ever to land on those worlds, they would need protection against the fiercely energetic particles, and the technology required to do that may be a long time coming.

The satellite deepest in the magnetic field, and the one therefore hardest to reach, would seem very familiar to us, if we could get to it, for it is a virtual twin of our own Moon. The satellite in question is Io (pronounced in two syllables, EYE-oh) and, of the four large satellites, is the closest to Jupiter.

Io's distance from Jupiter is 239,000 miles; compare that with the Moon's distance from the Earth, which is also 239,000 miles. Io's diameter is 2280 miles, just 5 percent greater than the Moon's diameter of 2160 miles.

Io's density, measured accurately now for the first time, is about 3.5 times that of water, one-third greater than had previously been thought. That means if you could weigh Io on a gigantic scale, you would find it would weigh 3.5 times as much as a sphere of water of the same size. The corresponding figure for our Moon is 3.4 times that of water. All told, then, Io has a mass only about 20 percent greater than that of the Moon.

This means that Io is an essentially rocky body, as our Moon is. No good view of Io's surface has yet been reported, but it wouldn't surprise anyone if it was cratered after the fashion of the Moon. Io and the Moon, then, are as near a case of twin worlds as we can find in the Solar system. Are there any differences?

Yes! The Moon, subjected to Earth's gravity, takes four weeks to complete its circle about Earth, moving at the speed of about 0.64 miles per second. Io, making a circle of exactly the same size around Jupiter, is made to move under the fierce lash of Jupiter's gravitational field (over 300 times as strong as Earth's) at a speed of 10.8 miles per second. It completes its orbit in 42.5 hours.

Then, too, Io's sky is much more spectacular than that

of the Moon in some ways. From the side of Io which eternally faces Jupiter, that giant planet would seem to hang motionless in the sky, while from the Moon we would see only Earth. Jupiter in Io's sky would be about ten times as wide as Earth would be in the Moon's sky, and over three times as bright.

Finally, Io is considerably colder than the Moon, since it receives only 1/25 the intensity of Sunshine that the Moon does. (Jupiter and its satellites are five times as far from the Sun as the Earth and the Moon are.) This means that even though Io has just about the same gravitational pull the Moon has, Io can manage to retain a tiny atmosphere, for gas molecules at Io's low temperature are too sluggish to escape easily. Pioneer 10 made note of Io's atmosphere, which thus became, for the first time, an observed fact.

The atmosphere isn't much, but it is considerably denser than that of our own Moon.

Io's atmosphere may consist of thin wisps of ammonia and methane, possibly a trace of water vapor, too. At its low temperature, any water or ammonia present would be mostly in the solid form as "ices." The ices would not collect at the poles as it does on Earth and Mars, for the distant Sun doesn't deliver enough heat to make the equatorial regions of Io much warmer than its poles. Instead, the various ices would lie in patches all over Io's surface.*

When Io moves behind Jupiter so that it is in the planet's shadow and receives no Sunlight at all (something that happens in the course of every revolution), the temperature drops a little and more of the atmosphere freezes out. When Io emerges from behind Jupiter and is bathed in Sunlight again, some of the ice evaporates. The density of the atmosphere moves up and down in a 42.5-hour cycle.

Of Jupiter's four large satellites, Io is the densest. As one moves outward, the satellites become less and less dense, something that Pioneer 10 has told us accurately for the first time.

Jupiter's second large satellite, Europa, has a density

---

* I was a little too self-assured here. It is quite possible for ice mixtures to be formed with a melting point low enough to be vaporized in the satellites' tropic zone and frozen at the polar zones. Pioneer 11, whose findings were published after this article was writen, took a photograph of Callisto that clearly shows a polar ice cap.

102

3.07 times that of water. Its third large satellite, Ganymede, has one that is 1.93 times that of water, and its fourth large satellite, Callisto, has one that is 1.65 times that of water.

It is as thought the cloud of dust and gas which enshrouded Jupiter when the planet and its satellites were forming was made up of lighter and lighter materials as one went outward from Jupiter.

Europa is the smallest of the four large satellites of Jupiter, only 1930 miles across, and has a mass only 0.6 times that of our Moon. Ganymede and Callisto, however, are enormous for satellites. Ganymede is 3490 miles across and Callisto is 3150 miles across, making them numbers 1 and 2 in size among all the satellites of the Solar system. Callisto is about the size of the planet Mercury, and Ganymede is considerably larger than that smallest of planets.

In terms of mass, Callisto is about 1.6 times as massive as the Moon and Ganymede is 2.5 times as massive. (Neither one is as massive as Mercury, however, for though Mercury is smaller than these giant satellites, it is considerably more dense; it squeezes more mass into its sphere.)

Since Io has a very thin atmosphere, the other three satellites ought to have one, too.

In one way, though, Ganymede and Callisto should be completely different from Io and from our Moon. Their densities are so low that they can't be made up chiefly of rock, as Io and the Moon are. They probably have rocky cores, but around those cores must be a thick layer of ices mixed with dark gravel.

As it happens, this is precisely the structure astronomers assign to comets—a mixture of gravel and various ices with a possible rocky core. It would seem, then, that the seven large satellites of the Solar system fall into two classes—rocky, or planet-like; and icy, or comet-like. In the first class are the Moon, Io, and probably Europa. In the second class are Ganymede, Callisto, and probably Titan, the largest satellite of Saturn. Concerning the seventh large satellite, Triton, which circles Neptune, we don't know.

Once a manned expedition reaches Mars, why not let Callisto be one of the next targets? It is the farthest of the four large satellites of Jupiter—over 1,170,000 miles from the planet—so it is least affected by Jupiter's giant gravity and least exposed to Jupiter's giant magnetic field. That means it would take less energy and involve less risk to reach Callisto and leave it again than would be true for the other large satellites.

Then, too, it would give us a chance to study a world that was completely different in composition and structure from any of the worlds of the inner Solar system. It would give us a chance to study something that may be best described as a giant comet.

AFTERWORD

There's no reason, incidentally, why the previous article might not have whetted your appetite for more information about Jupiter. In case it has, I might just mention that I have put out a book called Jupiter, the Largest Planet *(Lothrop, Lee & Shepard, 1973). I'd certainly have no objection to your buying it.*

13 THE NATURAL SATELLITES

FOREWORD

Nearly a year after I did the previous essay, the Encyclopaedia Britannica *asked me to turn out something for their* 1976 Yearbook of Science and the Future. *I suggested an article on the satellites of the Solar system and they agreed at once.*

There is some overlap in content between this article and the last, but the difference in audience, and therefore in style, tends to obscure the overlap somewhat, so I include both.

The natural satellites of the Solar system suffer by their association with the far more glamorous planets which they circle. The satellites are attendants, hangers-on, quickly dismissed in a final paragraph after a long discussion of the planet.

In recent years, however, that has changed. In the new age of space exploration, a number of the satellites are becoming interesting worlds in their own right.

This is true even for the one natural satellite that has

always seemed a distinct world to us by reason of its close-ness—our own Moon.

The Moon is, in many ways, the most remarkable of the satellites, quite apart from its relationship to ourselves. It is, of all the 33 known satellites of the Solar system, the closest to the Sun, since those planets closer to the Sun than Earth is, Venus and Mercury, have no satellites.

In relation to the world it circles, the Moon is by far the largest of all the satellites. Its diameter is 0.27 that of its primary, Earth, and its mass is 0.0123 that of Earth. No other satellite comes even close to this. In comparison to the planet it circles, the Moon is about ten times as massive as its closest rival. Earth and its Moon can reasonably be spoken of as a "double planet," something unique in the Solar system.

This has given rise to some speculation to the effect that the Moon is too large to be a true satellite of Earth; that it had once been a planet in its own right and had been captured by Earth.

True satellites that have formed in the same process that formed the planet, and that coalesced out of the outer edges of the dust cloud that condensed to form the planet, would be expected to revolve in the plane of their primary's equator and in the same direction as that of the planetary rotation. They would also be expected to have nearly circular orbits. These properties hold true for 20 of the 33 satellites.

Captured satellites would be expected to have eccentric orbits which are inclined, sometimes greatly, to the plane of the primary's equator. They would not necessarily be moving in the same direction as the planetary rotation, but might be "retrograde," moving in the opposite direction. They would generally be considerably farther from the planet than the true satellites are, and be smaller, too. Ten of the 33 satellites are generally accepted as having been captured. If the 20 undoubted true satellites are added to this, one can see that 3 of the satellites fall into a doubtful category. Among these doubtfuls is our Moon.

The eccentricity of the Moon's orbit is 0.055, a value greater than that for the undoubted true satellites, and its inclination to Earth's equatorial plane varies from 18° to 28°, and is only 5° from the plane of Earth's orbit about the Sun—a characteristic more appropriate to a planet than to a satellite. It is also unusually far from

Earth. Its distance of 380,000 kilometers is 60 times Earth's radius, and none of the true satellites is anywhere near that distance in terms of the radius of their primary. (This, however, may not be as important as it sounds, for the Moon is slowly retreating from Earth, and must have been considerably closer in ages past.)

Twenty-two kilograms of Lunar rock brought back by six successful flights to the Moon, between 1969 and 1972, show that the Moon's crust is at least 4.6 billion years old, indicating it to have reached its present state in the early period of the formation of the Solar system. The crust seems to lack water completely, however, and to differ from Earth's crust in being considerably poorer in those elements whose common compounds are volatile. That, combined with the common occurrence of glassy grains on the Lunar surface, would indicate that the Moon has experienced periods of surface heat of which there is no evidence in Earth's crust.

It may be that the Lunar crust was heated by the meteoric bombardments it suffered in the process of formation, an effect from which Earth's crust was protected by Earth's atmosphere and ocean, so that the differences are not inconsistent with Earth and Moon having been formed at the same time in approximately the same place and having remained together thereafter.

It may also be that the Moon, for some time after its formation (perhaps in Earth's neighborhood), was, for some reason, in an elliptical planetary orbit that carried it considerably closer to the Sun at perihelion, while leaving it near Earth's orbit at aphelion. The chief argument against this point of view is that the mechanics of capturing a body the size of the Moon by a body as small as Earth, and thus making a satellite out of a planet, involve conditions too constrained to make for easy credibility.

Nevertheless, if it ever becomes possible to analyze the crust of the planet Mercury, which is also airless and which is much closer to the Sun than Earth and Moon are, that might prove to have an important bearing on this question.

In one respect, the Moon is unique in Earth's sky. It is the only heavenly body that presents to us only one face. The far side of this nearest of objects was an impenetrable mystery until October 1959, when a Soviet Lunar probe passed beyond it and sent back the first crude photographs of that far side. Since then, more so-

107

phisticated probes have succeeded in mapping the entire surface of the Moon in great and accurate detail.

It turns out that the Moon has a surface asymmetry. The far side is heavily cratered but lacks the "maria," or "seas," that cover a large fraction of the side facing us. These seas, roughly circular, large, and relatively uncratered, were formed by catastrophic processes some half billion years after the rest of the Moon was formed. The seas may have been formed at a time when Earth's presence may have exerted an influence, perhaps for the first time—when the capture was effected.

Mars, the next planet in line from Earth, as we proceed outward from the Sun, has two satellites. These are insignificant bodies, however, so small as not to be discovered until 1877. Then, at a time when Mars was making a particularly close approach, the American astronomer Asaph Hall systematically scoured the neighborhood of the planet in search of possible satellites. He finally gave up, but his wife (maiden name, Angelina Stickney) urged him to try one more night. He did—and that was the night of the discovery.

The two satellites are named Phobos and Deimos ("fear" and "terror"), after the sons of the war god Ares (Mars) in the Greek mythology. Both move around Mars in nearly circular orbits in the equatorial plane and may be considered true satellites. They are remarkable for their closeness to their primary. Phobos, the inner satellite, is only 9350 kilometers from the Martian center, or 2.75 Martian radii. It is only 6000 kilometers above the Martian surface and its period of revolution is 7 hours and 39 minutes, the shortest for any of the 33 satellites. From the dimness of the two satellites, it was quite clear that they were very small bodies, and prior to 1971 nothing but their orbital characteristics were known.

In December 1971, the Mars probe Mariner 9 sent back photographs of the satellites and revealed them to be heavily cratered. This was the first time any satellite other than our Moon had been seen from a close distance.

In bodies as small as the Martian satellites, the feeble gravitation is insufficient to force matter into a spherical shape, and the crater impacts have clearly broken off bits and increased the irregularity. From the cratering, astronomers deduce that the satellites are solid rock and were formed early in the history of the Solar system, and were, perhaps, considerably larger bodies to begin with.

Phobos and Deimos are irregularly ovoid, with shapes remarkably like potatoes. Phobos has a long axis of 13.5 kilometers, an intermediate one of 11.5 kilometers, and a short one of 9.6 kilometers. The corresponding figures for Deimos are 7.5. 6.0, and 5.5 kilometers. (For comparison, the island of Manhattan is 20 kilometers long and 4 kilometers wide.)

Phobos has a volume of 5810 cubic kilometers and Deimos one of 1040 cubic kilometers. The volume of our single Moon is about 750,000 times as great as that of Phobos and Deimos put together.

Both Martian satellites have a low albedo and seem to be composed of a dark rock such as basalt. Each keeps the same face to Mars at all times, so that their periods of rotation are equal to their periods of revolution, as is the case with the Moon. For both Phobos and Deimos, the long axis points toward Mars, the intermediate axis is perpendicular to it and in the orbital plane, and the short axis is perependicular to both the others—in accord with what gravitational theory would predict.

From the photographs of Phobos a preliminary map of the satellite, showing some fifty craters, has been prepared and some of the craters have been named. One of the larger ones, near the south pole, was inevitably, named Hall. The largest of all, however, 8 kilometers across and on the side facing Mars, was named Stickney, after Hall's wife. The blow that formed Stickney seems to have produced a split in the rock that extends several kilometers eastward and which has been named Kepler Ridge, after the German astronomer Johannes Kepler, who first worked out the true orbit of Mars.

The most far-flung of all the satellite systems is that of Jupiter, the next planet beyond Mars. This is not surprising, since Jupiter is by far the largest and most massive of the planets.

Four of the Jovian satellites are large ones. These, reading outward from Jupiter, are Io, Europa, Ganymede, and Callisto, named for individuals with whom Zeus (Jupiter) had had love affairs in the Greek myths. They are collectively called the Galilean satellites, because they were discovered by the Italian astronomer

Galileo in 1610 (and were the first satellites. other than the Moon, to be discovered).

Their diameters, in kilometers, are, respectively, 3650, 2980, 5250, and 4900. Europa is somewhat smaller than the Moon, while the other three are larger. Ganymede, the largest of the Galileans, has one and a quarter times the volume of Mercury. Mercury, however, is the denser body and its *mass* (a more fundamental property than volume) is more than twice that of Ganymede.

From Earth, the Galilaean satellites can be seen as small disks in large telescopes, and faint markings have been used to show that the satellites keep one face to Jupiter as they revolve.

On December 4, 1973, the Jupiter probe Pioneer 10 took photographs of Ganymede from a distance of 750,-000 kilometers. These seem to show a mare over 750 kilometers across in the north polar region and another some 1300 kilometers across in the equatorial region. There also appear to be a few large craters.

Ganymede's effect on the path of Pioneer 11 allowed its mass to be calculated with unprecedented precision. From this and its volume, its density was shown to be about 1.8 grams per cubic centimeter.

This would indicate that Ganymede cannot be a ball of rock, since that would require a density of something over 3 grams per cubic centimeter (as is the case for Io and Europa, and for our Moon, too). Ganymede must contain sizable quantities of low-density frozen volatiles such as water ice, ammonia ice, and possibly methane ice. Callisto, which has a density of only 1.5 grams per cubic centimeter, must be even richer in such ices.

The Pioneer 10 photograph seems to show a bright region near Ganymede's south pole that could be an ice cap, which would be at least partly of methane. Methane (CH_4) is a gas at the ordinary temperature of the surface of the Galileans ($-145°C$.) but might freeze out during eclipses by Jupiter, particularly in the polar regions. Pioneer 11 sent back a photograph of Callisto, which shows an ice cap quite plainly.

The presence of volatile matter might indicate the possibility of an atmosphere, and traces of one that is about 1/20,000 as dense as that of Earth have been detected on Io. The other Galileans may have similar atmospheres.

Hydrogen has been detected near Io. It may have

been trapped within the body of the satellite in the course of its formation, and then have been slowly squeezed out to the surface. Io's gravity is not strong enough to hold the hydrogen, which, however, cannot escape from Jupiter's greater gravitational pull (even at Io's distance). The hydrogen remains in Io's orbit, therefore, and forms a kind of thin, gaseous torus, or doughnut, about Jupiter. Such a hydrogen torus has not been detected in connection with the other Galilean satellites, perhaps because the greater volatile content of the outer Galileans ties up the hydrogen as part of the ice molecules.

The density of the Galilean satellites increases as one moves in toward Jupiter. Perhaps, in the earliest days of the formation of Jupiter and its satellites, Jupiter was warm enough to heat the satellites significantly. Io, the closest, would have been baked driest.

Io is of an orange color and signs of sodium vapor have been detected in its thin atmosphere. It may be that the loss of the volatiles left Io's surface covered with a caked layer of impure sodium chloride (or dirty salt) and the impurities, whatever they may be, produce Io's color.

Pioneer 10 showed the magnetic field of Jupiter to be far larger than had been supposed. The Galilean satellites revolve within that field and this may, in years to come, limit or, at least, complicate, attempts to effect manned landings on them.

One Jovian satellite circles the planet within the orbit of Io. This is Jupiter-V or J-V (since it was the fifth satellite discovered—the discovery having been made in 1892 by the American astronomer E. E. Barnard). It is also called Amalthea, after the goat, or nymph, who nursed Zeus in his infancy. It is only 181,000 kilometers from Jupiter's center (2.5 Jupiter radii) and only 110,-000 kilometers above Jupiter's cloud layer. It is a small satellite, perhaps 160 kilometers in diameter, and circles the planet in 12 hours. The Galilean satellites and Amalthea are all true satellites, revolving in circular orbits in the plane of Jupiter's equator.

Beyond the Galilean satellites are eight more satellites, all smaller than Amalthea, with six of them being under 20 kilometers in diameter. All have high eccentricities and inclinations and are considered captured asteroids.

None of these outer satellites has been given an official name, but they are numbered in the order of discovery from J-VI (discovered in 1904) to J-XIII (which was discovered on September 14, 1974). The diameter of J-XIII may be no more than 8 kilometers (judging from its dimness), which would make it rival Deimos as the smallest of the satellites. They are sometimes given names of mythical characters associated with Zeus; the names (unofficial) in order of discovery from J-VI to J-XII are Hestia, Hera, Poseidon, Hades, Demeter, Pan, and Adrasteia. J-XIII has not yet even an unofficial name.

Of these outer satellites, J-VI, J-VII, J-X, and J-XIII form an inner group with orbits at an average distance of about 11 million kilometers from Jupiter. J-VIII, J-IX, J-XI and J-XII are an outer group with orbits at an average distance of 22 million kilometers. These four outermost revolve about Jupiter in retrograde fashion, and this is consistent with the theory that the outer satellites are captured asteroids, since capture at such a distance from a planet is easier in the case of retrograde motion than of direct motion.

J-VIII has the distinction of attaining a farther distance from its primary than is true of any other satellite. Its eccentric orbit carries it to a distance of 33,200,000 kilometers from Jupiter's center (465 Jupiter radii). It takes J-VIII 735 days to complete a revolution about Jupiter. J-IX, which has a slightly greater average distance, completes its revolution in 758 days (2.07 years) and has the longest period of revolution of any of the 33 satellites.

Beyond Jupiter lies Saturn, which is second only to Jupiter in size and mass and, therefore not surprisingly, is second only to Jupiter in the number of its satellites. It has ten satellites with names, counting outward from Saturn, that are Janus, Mimas, Enceladus, Tethys, Dione, Rhea, Titan, Hyperion, Iapetus, and Phoebe. Of these, all but Janus are named for giants or Titans who, like Cronus (Saturn) himself, fought against Zeus in the Greek myths.

The largest Saturnian satellite is Titan, which was discovered in 1655 by the Dutch astronomer Christian Huygens. Until very recently, Titan was thought to be somewhat smaller than Ganymede and Callisto. On March 30, 1974, however, the Moon passed before it, and from the speed of the Moon's motion and the time it took to cover Titan, it was found that Titan's diameter was 5800 kilometers. This unexpectedly large value has caused Titan to be recognized as the most voluminous satellite in the Solar system.

Titan's low density, however, causes it to have a smaller mass than Ganymede. The latter, with 2.1 times the mass of our Moon, still retains its title as the most massive satellite in the Solar system.

The low density of Titan is significant, because it means that, like the outer Galileans, it is composed largely of volatiles. Its temperature is lower than that of the Galileans and its gravitational pull can better hold the molecules of methane gas, made sluggish by the cold.

In 1943, methane was detected in the neighborhood of Titan, in considerable quantity. This makes Titan unique among satellites as the only one of the 33 known to possess an atmosphere in more than traces. Its atmosphere is almost surely thicker than that of Mars and it may approach the thickness of Earth's atmosphere. Titan is also the origin of hydrogen gas which it cannot hold and which forms a torus around Saturn in Titan's orbit, as Io's hydrogen forms a torus around Jupiter in Io's orbit. Like Io, Titan is orange in color.

Saturn's nine other satellites are of moderate size. The smallest is the outermost, Phoebe, which has a diameter of about 200 kilometers. Its average distance from Saturn is 13,000,000 kilometers, nearly four times as great as that of Iapetus, next outermost. Its great distance, the eccentricity of its orbit, and the fact that it revolves in retrograde fashion make it seem certain that Phoebe is a captured asteroid.

Iapetus, which may be as much as 1750 kilometers in diameter, is the second-largest of the Saturnian satellites. It has an orbit inclined to the plane of Saturn's equator by 14.7°, which makes it uncertain (as in the case of our Moon) whether it is a true satellite or a captured body. Iapetus is six times as bright when it is west of Saturn, when it exposes one hemisphere to us, as when it is east of Saturn and exposes the other (assuming it always turns the same face to Saturn as it revolves). It is not known why this is so, but perhaps one hemisphere is predominantly icy, the other predominantly rocky.

The inner eight Saturnian satellites rotate in the plane of Saturn's equator and would seem to be true satellites,

113

although the orbits of Titan and Hyperion are rather eccentric.

The most recently discovered of the Saturnian satellites is the closest, Janus. It is only 157,000 kilometers from Saturn's center (2.6 planetary radii) and completes its rotation in 18 hours. Because it is the first of the satellites in order, but the latest discovered, it was named for the Roman god of beginnings and endings.

Within Janus's orbit are Saturn's rings, flat, wide objects that encircle Saturn in its equatorial plane. The extreme width of the rings, measuring across Saturn, is 270,000 kilometers, but they are not more than 15 kilometers thick.

We see the rings at various angles in the course of Saturn's 29-year period of revolution about the Sun. Twice each Saturnian revolution, we see them edge-on, when they are so thin as to disappear. It is when they disappear and their brightness is eliminated that the inner Saturnian satellites are most easily visible. It was when the rings were edge-on in 1966 that Janus was discovered. Janus's orbit is only 21,000 kilometers beyond the outermost edge of the rings.

The rings are not solid but are a collection of innumerable particles, each of which may be considered a separate satellite of Saturn, which, from that point of view, has innumerable satellites. There are no sharp boundaries to the rings, but those regions where the swarm is still thick enough to see from Earth extend inward to about 75,000 kilometers from the center of Saturn (1.25 radii) or only about 14,500 kilometers above Saturn's cloud layer. The innermost particles of Saturn's rings rotate about Saturn in 8 hours.

None of the 33 satellites of the Solar system is as close to its primary in terms of the planetary radius as are the innermost particles of Saturn's rings.

The planet Uranus has a compact and, on the whole, unremarkable family of five satellites. Their names, reading outward from the planet, are Miranda, Umbriel, Ariel, Titania, and Oberon, and these are *not* drawn from Greek mythology. They are, rather, the names of characters in the works of Shakespeare and Pope.

Of the four outer planets, Uranus is the only one that does not have a really large satellite. The largest is Titania which has a diameter of about 1100 kilometers. Uranus is also the only outer planet without at least one captured satellite. All five of the Uranian satellites revolve in the equatorial plane of the planet in orbits of virtually zero eccentricity.

To be sure, the five satellites all revolve in a plane that is nearly at right angles to the plane of Uranus' revolution about the Sun, and no other planet has satellites of which this can be said. However, this is because Uranus' equatorial plane is at nearly right angles to the plane of its revolution, and the peculiarity is rather that of the planet than of its satellites.

Neptune has two satellites, and each is, in its own way, remarkable.

The outer satellite, Nereid, named for the ocean nymphs who were the daughters of Poseidon (Neptune) in the Greek myths, has the earmarks of a captured satellite. It is small, perhaps 300 kilometers in diameter, and has an eccentric orbit inclined almost 28° to the plane of Neptune's equator. Its eccentricity is 0.76, higher than that of any object in the Solar system, other than comets. This means that although the average distance of Nereid from Neptune is 5,550,000 kilometers, the orbit is so elliptical that at one end it comes to within 1,400,000 kilometers of the planet and at the other recedes to a distance of 9,500,000 kilometers.

The inner satellite, Triton, named for the son of Poseidon, is a large satellite, 3700 kilometers in diameter, a little larger than the Moon, and larger in comparison with its primary than any satellite but the Moon. Its distance from Neptune is 355,000 kilometers, just a little less than the distance of the Moon from Earth, and unlike the Moon, Triton's orbit is nearly perfectly circular.

So far, Triton has the earmarks of a true satellite, but the plane of its orbit is at an angle of some 20° to Neptune's equatorial plane, about equal to the tipping of the Moon's orbit. And there is worse, too. The Moon, at least, revolves in the same direction as Earth's spin. Triton's motion is retrograde, in the direction opposed to Neptune's spin.

If Triton, like a true satellite, was formed on the fringes of the Neptune dust cloud, it should revolve in the direction of Neptune's spin. If, on the other hand, it is a captured satellite, as one might think it would have to be if it revolves in retrograde fashion, then what impossibly lucky stroke put it into an almost exactly circular orbit?

It is one more satellite mystery to be plumbed someday
—we can only hope.

AFTERWORD

I would like to point out, by the way, the rapidly increasing interest in what we might call "futurism." It was not so long ago that encyclopedias put out what were simply yearbooks of "science." Now the Britannica at least, recognizing the strong link between science and the manner in which the future may be molded, makes it "science and the future."

I approve of this, naturally. I am, after all, a professional futurist and much of my writing, both fictional and nonfictional, deals with the future.

14 OF LIFE BEYOND

FOREWORD

Back in the summer of 1974, James L. Christian, a professor of philosophy at Santa Ana College in California, was planning a collection of essays dealing with the effect of the discovery of extraterrestrial intelligence on various aspects of human culture. He asked me to write a general article on the history of human speculation concerning extraterrestrial life and I responded with "Of Life Beyond."

He had offered a reasonable payment for the article and delivered said payment in the form of a personal check. This made me uneasy, and when his next letter displayed, between the lines, a certain dubiousness concerning the fate of the book, I wrote to ask if he had a publisher.

It turned out he hadn't and that he had paid for the article out of his own pocket. My uneasiness rose to new heights, for I saw clearly that unless he found a publisher — I would be forced to return the money. While I am sufficiently full of integrity to make the return. I am not sufficiently full to enjoy doing so.

Feverishly, then, I began to try to find a publisher for Jim, and I wish I could say that I turned the trick. I didn't, however, and when I was about to give up hope, Jim Christian found one entirely without my help. The book appeared in the summer of 1976. It was entitled Extra-Terrestrial Intelligence: The First Encounter, *was extremely good-looking, had first-rate articles in it—and I did not have to return the money.*

If we are going to speculate about extraterrestrial life, we must first ask what we mean by the phrase.

We can define "terrestrial life" without difficulty. That would include the millions of species of plants, animals, and microorganisms that now exist on Earth, or have existed on Earth in the past. We can then define extraterrestrial life as all forms of life that do not now exist on Earth and never have existed on Earth.

If that be so, then almost all human beings who have ever lived on Earth, including the vast majority of those who live on Earth now, firmly believed, or believe, in the existence of extraterrestrial life of any, or all, of a number of kinds.

There are, to begin with, life forms that would fit into the natural scheme of things perfectly well, if they only existed. These would include unicorns, dragons, rocs, and similar creatures. They possess only animal intelligence, though they are more beautiful or more dreadful than the animals well known to man.

Then there are various kinds of para-natural life forms, those which follow laws of behavior radically different from those we know and which possess abilities, very often, far beyond human experience. Usually, such life forms are marked by human or superhuman intelligence.

There are life forms of this sort that are more or less neutral in their attitude toward humanity. There are talking animals, for instance. There are the elves, fairies, gnomes, and "little people," generally with attitudes toward humanity varying from the slyly mischievous to the mildly benevolent. There are also the various ghosts and spirits of the dead. If the readers of this article do not accept the existence of such life forms, they should recognize themselves to be a rather inconsiderable minority of the human race even now.

Among the actually superhuman intelligences that have

been widely accepted both in the past and in the present, too, there are those that are malignant and that are consumed with evil intentions toward humanity—the entire tribe of ogres, afrits, demons, right down (in our own particular culture) to the Devil himself.

Finally, there are the superhuman intelligences that guide the Universe and that, more often than not, are viewed as vastly benevolent; all the divine and semi-divine creatures of legend, up to (in our own culture) the angels, archangels, and God Himself.

Yet all of these we can eliminate from consideration here. Whether any of them exists or not is irrelevant. When we speak of extraterrestrial life today, we are referring to those forms of life which are bound by the same laws of the Universe that bind us, but which differ from the life we know only in that they occur, not on Earth, but in some other part of the Universe.

It might be more accurate to speak of extraterrestrial life as "otherworldly life" were it not for the fact that if we conceived of life existing in dust clouds filling the vast spaces between the worlds, that, too, would be a form of extraterrestrial life.

If we come down, then, to extraterrestrial life in the astronomical sense, it is clear that there could be no sensible speculations concerning such things until the time came when human beings understood that there were indeed worlds other than the Earth.

This is by no means self-evident. In the earliest legends, the whole visible Universe was conceived of as a relatively small patch of flat Earth over which was draped a solid firmament, coming down to the horizon on all sides. In the firmament, the stars were mere luminous pinpoints and the Sun and Moon were lamps with no other function but to light the Earth, and were probably not much larger than ordinary lamps.

The Earth, by this view, was all the habitable Universe there was, and nothing could lie beyond but the para-natural abodes of gods or demons.

What upset this limited picture of the Universe was the Moon. All the other heavenly bodies seemed to be pure light, unchanging and eternal. The Moon, however, had visible shadowings on its face that spoiled its perfection. What's more, the Moon changed shape, from a perfect circle of light to half a circle, then to a crescent, then back again.

Anyone studying the changing position of the Moon relative to the Sun during the course of these phase changes was bound to see that the lighted portion of the Moon always faced the Sun. From this one could conclude that the Moon was a dark body, which shone only by the reflected light of the Sun, and that the phase changes were the inevitable consequence of the movements of Moon and Sun across the sky. The mere fact that the Moon was intrinsically dark gave it at least one property in common with the Earth, marked it with a world-like aura, and produced the first hint of the multiplicity of worlds.

As for the dim markings on the Moon, it was inevitable that, given the anthropocentrism of humanity, it would be seen as the barely made out form of a living human being on the Moon. It is this sort of thing that gave rise to the legends, in our own culture, of "the Man in the Moon," supposedly the biblical Sabbath-breaker of Numbers 15:32-35.

The Man in the Moon legends may represent the first speculation concerning extraterrestrial life in the modern sense.

Yet though the Moon was a dark world like the Earth, it could, of course, be pictured as no larger than it seemed to be, so that the Man in the Moon filled the world he lived on.

About 130 B.C., however, a Greek astronomer, Hipparchus of Nicaea, calculated the distance of the Moon accurately for the first time, and found it to be thirty times the Earth's diameter. Combining this with the best Greek value of the Earth's diameter, the Moon's distance from Earth turned out to be 240,000 miles. To appear as large as it does in the sky from that distance, the Moon must be some 2000 miles across.

This was the first definite indication in human history that at least one other world existed in the Universe, a real world with dimensions that were a respectable fraction of Earth's size.

If that were so, what of the other planets, whose distances could not be determined by the ancient astronomers, but which were known to be farther off than the Moon? Might they, too, be worlds like the Earth and the Moon?

The conclusion was a fair one and the lively imagination of writers could seize upon that thought. In the second century A.D., the Syrian writer Lucian of Samosata wrote the first interplanetary romance that we know of. He tells of a ship that was carried up to the Moon by a waterspout. He describes the intelligent beings living on the Moon and tells of the war they are conducting against the intelligent beings of the Sun over their conflicting ambitions to colonize the world of Venus.

What seemed true, indirectly, from Hipparachus' measurement of the Moon's distance was backed by more direct evidence in 1609, when the Italian scientist Galileo Galilei turned a telescope on the heavens for the first time. When he looked at the Moon, he saw a world on which there were mountains and craters. He also saw flat regions he called seas. The Moon was visibly a world like Earth in these respects at least.

When Galileo turned his telescope on the planets, he saw them magnified into small globes—small, it was obvious, only because of distance. The Moon and the planets were worlds beyond dispute and this gave a new impetus to dreams of extraterrestrial life.

An English clergyman, Francis Godwin, wrote a story called *Man in the Moone,* published in 1638. Godwin's hero flew to the Moon in a chariot hitched to great geese which were supposed to migrate to the Moon regularly. Godwin described the moon as a world much like the Earth, but better.

This book was the first of the line of modern interplanetary romances, a line which has continued and flourished to this day, and which still forms a major theme in the fabric of contemporary science fiction.

In general, those who thought of other worlds invariably imagined life upon them; almost always intelligent and rather man-like life. There seemed the unspoken assumption that God would not waste a world. If worlds existed, their only purpose had to be to bear man-like beings.

Nor was it only romancers who thought so. Consider the German-English astronomer William Herschel, who in the decades immediately preceding and succeeding 1800 was the foremost astronomer in the world. Herschel could not even bear to spare the Sun. He suggested that the Sunspots were holes in the flaming atmosphere of the Sun; that through them the dark and cool surface of the Sun's globe itself could be seen; and that this dark and cool surface might be inhabited.

Even as late as 1935, the New York *Sun* was able to run a series of hoax stories about a Moon on which there were Earth-like conditions and intelligent life, and found (perhaps to its surprise) that the general public believed it. The *Sun's* circulation boomed till the hoax was exposed.

Yet it was not long after Galileo's initial observations of the Moon by telescope that it became rather obvious that the Moon's seas were not of water but of flat, dry land and there were no signs of either water or air upon our satellite. There were no clouds on the Moon, no twilight, no change of any kind. It was an airless, waterless world.

Of course, it did not follow, inevitably, from the fact that no air or water could be detected on the Moon that it was without life. Yet if there were life on the Moon, it would have to be radically different from our own and there was no evidence that such "other-life" existed. Among astronomers, it grew steadily more common to view the Moon as lifeless.

It was thus borne in upon mankind, for the first time in history, that it was possible to have a dead and barren world.

The first interplanetary romance that tried to take into account the state of scientific knowledge of the time (and was therefore perhaps the first real science fiction story) was written by the German astonomer Johannes Kepler. It was published posthumously in 1634.

In this tale, *Somnium,* the hero is transported to the Moon in a dream. There he found a day and night that were each two weeks long (something which even primitive stargazers knew, but a fact which romancers almost invariably ignored). Kepler could not abandon life, however. He postulated strange animals and plants that grew rapidly during the long day and died at nightfall.

In 1643, another blow at romance was struck by the Italian physicist Evangelista Torricelli. In that year, he invented the barometer and showed the pressure of the atmosphere to be equal to that of a column of mercury thirty inches high.

If the atmosphere exerted only that great a pressure, it would have to extend upward no more than five miles, assuming it to be the same density all the way up. In

122

actual fact, the density of the air decreases rapidly with height, so that the atmosphere extends much higher than five miles, but it rapidly grows too thin to support life.

The universal assumption of earlier times that air extended indefinitely upward (so that men could travel to the Moon in chariots hitched to flying geese) was destroyed. It became clear that the various worlds of the Universe were separated by vast stretches of vacuum which acted as an insulating layer as far as life was concerned. If there were life on other worlds, reaching them or having them reach us would be a matter of enormous difficulty.

Oddly enough, the first suggestion of the one possible means of reaching the Moon through the vacuum beyond the Earth's atmosphere came not from any man of science, but from a science fiction writer—none other than Cyrano de Bergerac. Cyrano, the long-nosed duelist, really existed, really had a long nose, really fought duels, and was also a clever writer. In 1650, he published a book called *A Voyage to the Moon*, which treated the Moon as a world inhabited by intelligent beings. In the course of the book, he suggested several methods of reaching the Moon, and one was to tie rockets to a chariot, light them, and zoom off.

To this day, the rocket engine is the one propulsive method known to man by which it is possible to cross a stretch of vacuum and steer a course through space. It is the method by which mankind finally (three centuries after Cyrano's tale) reached the Moon.

By the nineteenth century, the Solar system was found to contain many more worlds than had been known in the days before the telescope. A seventh planet, Uranus, had been discovered in 1781, and an eighth, Neptune, in 1843. It was known that the outer planets were far larger than Earth, and that each had a train of satellites which, taken together, represented some dozen and a half additional worlds of varying size. There were numerous small asteroids circling the Sun in the space between the orbits of Mars and Jupiter.

However, nineteenth-century astronomy had grown quite sophisticated, and as more and more was known about these various worlds, the outlook for life grew more and more gloomy. In the 1860s, the Scottish mathematician James Clerk Maxwell worked out the molecular interpretation of the gas laws and it came to be understood why the Moon was airless. Its gravitational field was insufficiently intense to hold the rapidly moving molecules of gases to its surface. It would seem from this that any world as small as the Moon, or smaller, would have no atmosphere, or, at best, one too thin to support our kind of life.

All the satellites and asteroids were therefore airless or nearly so. Mercury was larger than the Moon, but also closer to the Sun, and therefore much hotter and that meant it, too, was airless.

The larger outer planets did have atmospheres, but those planets were so large that their gravitational fields were intense enough to collect and maintain atmospheres of such enormous pressures as to make their environments utterly un-Earth-like.

Again, we can't entirely eliminate the chance of strange life forms capable of enduring conditions we would consider intolerable, but there is not even the faintest evidence for the existence of any life of a nature wildly different from our own.

Some have indeed speculated on the kinds of chemistry that might form the basis of life on worlds much colder than our own, or much warmer, or with radically different atmospheres and oceans. There have been thoughts of ammonia playing the role of water on cold planets, while silicones or sulfur did so on hot planets, and complex silicates built themselves into life forms on liquid-free planets. Such speculations, barren of any observed evidence at all, lead nowhere unfortunately and on the whole astronomers have found it more useful to restrict themselves to the possibility of extraterrestrial life similar in basic chemistry to our own.

The question was, then, whether any world could or did possess the type of environment to which life similar to the life forms on Earth could possibly adapt itself.

By the second half of the nineteenth century, the possibilities of this, among the known worlds at least, reduced themselves to two, and two only, outside the Earth. These were the planets Venus and Mars.

Venus is almost as large as Earth and is closer to the Sun. It might be warmer, but perhaps not too warm for life. It obviously had an atmosphere and on it was borne a permanent layer of clouds that should protect the surface from the too ardent embrace of the Sun. It was easy, then, to speculate that Venus was a watery primitive world that might even have a planet-girdling ocean. However, since there was no way of peering through the eternal cloud layer at the surface below, there seemed no way of advancing beyond that speculation.

As for Mars, it is smaller than Earth and has a thinner atmosphere, yet perhaps not too thin. It is farther from the Sun than Earth is, and is therefore colder, but perhaps it is not too cold. Mar's axis of rotation is tipped to its plane of revolution about the Sun by almost exactly the amount Earth's axis is tipped. This means that Mars has Earth-type seasons (though colder). It has ice caps at either pole that expand and contract with those seasons, and color changes that seem to represent advancing and retreating vegetation.

Mars is unique in another respect. Of the worlds beyond the Moon, only Mars is close enough and has an atmosphere clear enough to allow its surface to be mapped. Naturally, the best map would be produced when Mars makes a particularly close approach to Earth.

Every two years or so, Earth passes Mars in their two orbital motions about the Sun. These points of passing ("oppositions") come at different places in the planetary orbits, orbits which are closer to each other in some places than in others. In 1877, Earth was in opposition to Mars at points in their orbits which were nearly at minimum separation. Mars would then be seen with the kind of clarity that came only three times a century, and many astronomers planned to study it then.

One of those who studied Mars at the 1877 opposition was the Italian astronomer Giovanni Virginio Schiaparelli, He produced the first modern map of Mars, one that persisted essentially unchanged until the development of techniques for observing Mars that proved far more powerful than the telescope.

In drawing his map, Schiaparelli noted dark areas, which he considered to be bodies of water, and light areas, which he considered to be land. Connecting the dark areas, Schiaparelli saw a few markings that were rather straight and thin. He thought these were narrow connecting waterways, rather like the various straits and channels on Earth. Indeed he called them "canali," which is the Italian equivalent of "channels" and merely means narrow waterways.

Unfortunately, "canali" was translated into English as "canals," which are, of course, artificial waterways. The notion arose and grew stronger that Mars, with its gravitational field at its surface only two-fifths that at Earth's surface, was gradually losing its water; that an ancient dying civilization on that planet was intent on preserving what water still existed as long as possible and making as efficient use of it as possible. The canals, it was thought, were an advanced engineering plan to keep the planet irrigated.

For the first time, it seemed that science had produced direct evidence not only for life on another world but for *intelligent* life and a highly developed civilization.

In 1894, the American astronomer Percival Lowell established an observatory in Arizona. For years he studied Mars closely through the clear air of the southwestern desert and drew maps showing an intricate lacework of canals. Eventually he placed five hundred of them on his maps. He also plotted the "oases" at which they met, recorded the fashion in which the canals seemed to double at times, and noted in detail the seasonal changes that seemed to mark the ebb and flow of agriculture. Lowell, who in 1908 published a book entitled *Mars as the Abode of Life,* was the foremost proponent of life on Mars.

Even more influential than Lowell in convincing people generally that there was intelligent life on Mars was the English science fiction writer Herbert George Wells, who in 1898 responded to the growing popularity of the Schiaparelli-Lowell view by publishing *The War of the Worlds.*

In this book, Wells told of Martians who, despairing of being able to maintain their dying world, launched an invasion of Earth. Their superior technology made it possible for them to overwhelm the inhabitants of the Earth, but the Martians were nevertheless defeated by physiology, for their bodies could not resist the onslaught of Earth's decay bacteria. This was the first story to deal with interplanetary warfare, and it introduced the first note of fear into speculation about extraterrestrial life.

The belief that there was a civilization on Mars remained popular with the general public well into the mid-twentieth century. As late as 1938, when a realistic version of *The War of the Worlds* was placed on American radio by Orson Welles, and Martian ships were described as landing in New Jersey, there was panic among parts of the population there.

Astronomers, however, did not all agree on the Martian canals by any means. Many excellent observers did not succeed in seeing canals on Mars. The feeling arose among some of them that the canals were an optical illusion; that there were irregular blotches on the Martian surface which, individually, were just below the limits of vision, but which, taken together, were interpreted by straining eyes as straight lines.

There was no way of deciding between those who supported canals and those who denied them as long as only telescopic views were possible, but throughout the twentieth century, as more and more was learned about Mars, the less and less hospitable the world seemed to be. Mars was colder and drier than had been thought. The atmosphere was thinner and contained no oxygen. Even the ice caps might be frozen carbon dioxide rather than water.

Finally, in 1965, the Mars probe Mariner 4 took photographs of the planet Mars from a distance of 6000 miles above its surface. The photographs showed many craters, but no canals. Still more sophisticated probes since then have succeeded in mapping the entire Martian surface in detail. There are no canals. There are giant volcanoes and canyons and many other fascinating features, but no canals. What Lowell saw were indeed optical illusions.

It may still be that there are simple forms of life on Mars, bacteria-like or lichen-like forms. It may be that Mars suffers alternations of climate and that we are now viewing it in its "ice age" period. Under other conditions its environment may be much milder and more hospitable to life.

A decision as to that will have to await a manned landing on Mars, perhaps but a dispassionate assessment of the situation must make it appear that the chance of finding life on Mars is rather low, while that of finding *intelligent* life is virtually nil.*

* Since this article was written, Vikings 1 and 2 have both landed safely on Mars and have sent back beautiful photographs of the Martian surface. Experiments designed to detect life have been carried through and the results have been ambiguous. The chemical activity of the soil is surprisingly high and *may* be due to the presence of life, but on the other hand, none of the organic compounds characteristic of life as we know it can be detected. The puzzle remains and it is more vital than ever to explore Mars further. Biological or chemical, that soil activity has knowledge to impart to us.

In 1962, meanwhile, the Venus probe Mariner 2 confirmed what astronomers had been suspecting for several years—that Venus was unexpectedly and extraordinarily hot. Its thick atmosphere, a hundred times as dense as our own, was almost entirely carbon dioxide and such an atmosphere trapped heat (the "green-house effect") and raised Venus's surface temperature to over 400° C. Far from being a waterlogged planet, Venus is incredibly hot and bone dry. Life as we know it cannot possibly exist on Venus.

Finally, in 1969, human beings stood upon the soil of another world—the Moon—for the first time, and found it to be, as expected, completely lifeless.

So it would seem at the moment of writing that there is only one world in the Solar system, outside Earth itself, on which there may be life. That world is Mars. and even there any life that exists will probably be very simple—and the chance of any life at all existing is not great.

We have no choice but to admit that the odds are rather heavily in favor of the Solar system (outside Earth) being dead.

There is, however, a vast Universe outside the Solar system; what of that?

Beyond the Solar system are the stars, but they were for a long time discounted. In early times, when the Moon and Sun seemed worlds, and when it was already possible to assume the planets might be worlds, too, the stars were still viewed as merely decorative markings on the solid curve of the sky.

The first person we know of who thought of stars as worlds was a German cardinal, Nicholas of Cusa, who in 1440 published some remarkably modern-sounding notions of the Universe. He held that space was infinite and that the stars were other suns. Since it seemed unnatural to him that all those suns would be wasted, he assumed that each had its family of planets circling it, and that those planets were inhabited.

The Church was secure in 1440 and Nicholas of Cusa did not get into trouble over his views. A century and a half later, the Italian philosopher Giordano Bruno vehemently upheld views similar to those of Nicholas, but this was at a time when religious disputes were racking Europe. In 1600, Bruno was burned at the stake for his various heresies.

The coming of the telescope, and the revelation that the outer planets were hundreds of millions of miles distant, made it clear that the stars, which had to be farther off still, must be extraordinarily luminous to be visible at all. More and more astronomers began to think of them as distant suns.

In 1718, the English astronomer Edmund Halley reported that some of the brighter (and, therefore, possibly nearer) stars, such as Sirius, Procyon, and Arcturus, had shifted position since ancient times. This was a demonstration that the stars were not fixed to the sky, but were freely moving like individual bees in a swarm. If they appeared motionless it was because their enormous distances made their motions seem very slow. Halley's discovery put a final end to any notion that the stars might be anything but suns.

It is not surprising that the French satirist Voltaire, in a story called *Micromégas,* published in 1752 and describing two extra-terrestrial visitors, had one of them come from Sirius.

It was not till 1838, however, that the German astronomer Friedrich Wilhelm Bessel became the first to measure the distance of a star. The star he chose to study was a rather dim binary star (consisting of two stars close together and circling each other) named 61 Cygni. He chose it because it moved against the background of the stars more rapidly than any other star was known to—thus indicating it might be unusually close. Checking its slight shift in position as the Earth moved around the Sun, Bessel estimated 61 Cygni to be some sixty million million miles away.

Although light travels at the speed of 186,000 miles per second, it takes light eleven years to cover this great distance. The star, 61 Cygni, is therefore eleven "light-years" away. Even the nearest of all stars proved to be a little over 4 light-years away.

In the century that followed, the known size of the Universe constantly increased. The stars we see in the sky are all part of a system called the Galaxy, and the Milky Way is what we see when we look through the long axis of the Galaxy—endless numbers of faint stars. Actually, it is estimated there may be 135,000,000,000 stars in the Galaxy, and there may be as many as 100,000,000,000 other galaxies distributed through space.

The total number of stars in the Universe is quite beyond comprehension, but so are the distances that separate them.

Nor can we conquer those distances by building up equally vast velocities for our spaceships. In 1905, the German scientist Albert Einstein demonstrated that the speed of light in a vacuum was the fastest speed that massive objects could ever attain. Any probe traveling to even the nearest star at even the greatest conceivable speed would have to take 4.3 years for the journey.

It seemed discouraging to have to speculate about extraterrestial life in any worlds located at the vast distances of the stars. As long as it seemed that there might be extraterrestrial life in the Solar system, then, astronomers wasted little time speculating about the stars.

One route by which stars entered into speculations dealing with the planets involved the question of the origin of planetary systems.

In 1798, the French astronomer Pierre Simon de Laplace had suggested that the Solar system was originally a large cloud of gas and dust (a "nebula") that whirled majestically, and slowly condensed into the Sun. As it did so, it gave off small portions of itself that formed the planets.

Over the course of the nineteenth century, however, Laplace's "nebular hypothesis" crumbled under the weight of various objections. It was hard to see, for instance, how enough of the whirling property ("angular momentum") of the cloud could be concentrated into the small portions that formed the planets. Some 98 percent of the angular momentum of the Solar system is concentrated into the planets, which, taken all together, are only one-thousandth as massive as the Sun.

In 1917, therefore, the English astronomer James Hopwood Jeans suggested an alternative explanation. He proposed that the Sun and some other star had had a near collision and that the gravitational pull of each had served to extract matter from the other. This extracted matter was given a strong spin by the gravitational fields of the stars, as they moved apart.

By this theory, which was widely accepted for over a quarter of a century. planetary systems arose only when two stars passed close to each other. Considering the distances that separate the stars and the speed with which they move, such a near collision must be excessively rare.

130

In the entire history of the Galaxy, it might have happened only once.

By Jeans's theory, then, there would possibly be only two planetary systems in our Galaxy; our own and that of the star that passed close to the Sun. And it might be that in these two planetary systems, our own planet is the only one that bears life. The vision to which Jeans's theory gave rise, then, was of a glorious Universe, utterly free of life except on the Earth and at most on one or two other such planets per galaxy.

Jeans's theory had shortcomings, too, and these grew to seem more serious as astronomers' knowledge of the inner structure of stars increased. In 1944, the German astronomer Carl Friedrich von Weizsäcker suggested a return to the theory of Laplace at a more sophisticated level. Turbulence of the original dust and gas was taken into account and, in later modifications of the theory, electromagnetic forces as well as gravitational ones.

If Weizsäcker's theory is correct, then planetary development is a normal step in the birth of a star, and nearly every star ought to have planets circling it.

Even as Weizsäcker was proposing his theory, observational evidence was supporting it. In 1943, the Dutch-American astronomer Peter Van de Kamp reported small irregularities in the movements of one of the two stars of 61 Cygni. These small irregularities, it could be deduced, were caused by the gravitational pull of a large planet eight times the mass of Jupiter. In 1960, a planet of similar size was shown by Van de Kamp to be circling the small star Lalande 21185, and in 1963, a smaller planet (possibly two) seemed to be producing irregularities in the motion of Barnard's Star.

Stars are so distant that it takes a particular set of conditions to make planets detectable. The stars must be particularly small and the planets particularly large, so that the gravitational effects of the latter on the former are considerable. As it happens, Barnard's Star is second-closest to us, Lalande 21185 third-closest, and 61 Cygni twelfth-closest. That three planetary systems should be detected in our immediate neighborhood, even though the requirements for detection are stringent, is extremely unlikely unless planetary systems are common indeed.

At the present time, therefore, astronomers are generally convinced that planets are common phenomena and that most or all of the stars in the Universe possess some sort of planetary system, and that there are therefore innumerable worlds on which life might conceivably develop.

By the mid-twentieth century, of course, scientists were no longer ready to assume that any world would have life on it as a matter of course. There was no longer the feeling that a world without life was a world wasted; experience with the Solar system itself had revealed how easy it was for a planet to present an environment so basically different from that of Earth that life as we know it was not in the least likely to exist.

It was legitimate to wonder, then, if worlds might not remain barren even though they seem, to our eyes, to be potentially hospitable to life. Perhaps, even though planets are a common and even inevitable phenomenon, life is not. Life might have its origin only as the result of an incredibly rare chance, and the Universe, though it might be studded with likely planets, might be empty of life except for that on our own world.

A definite impression that life might not be the result of a rare accident came about with a line of experiments that began in 1952 with the work of the American chemist Stanley Lloyd Miller. Beginning with a sterile mixture of simple compounds of the type that might have existed on the primordial Earth— hydrogen, methane, ammonia, and water—Miller added energy in the form of an electric discharge, and in a week's time found that more complicated compounds had been formed, including two of the amino acids that form the building blocks of protein molecules.

Later experiments by others made it quite clear that simple compounds plus energy yielded complex compounds that were invariably of types that seemed to point in the direction of life.

Nor was this apparent only in the laboratory. Beginning in 1968, signs of atom combinations were discovered in the dust clouds found in interstellar space. Through their absorption of specific radio wavelengths, more than a score of compounds revealed themselves to astronomers, several of these, too, pointing in the direction of life.

It seems quite certain, then, that life is the result of a very natural tendency—an obvious route taken by compounds under the whip of energy toward a steadily growing complexity. We might suppose that had a different route been the natural and obvious one, life would have begun in another fashion, based on another kind of chemistry—but one route to complexity was *the* route and it gave rise to *our* kind of life.

Since life would seem not to be the result of a rare accident, and since life, approximately as we know it, would seem to be on the natural line of development for any planet similar in mass, temperature, and chemistry to our own, some astronomers have begun to calculate probabilities.

They estimate the number of stars per galaxy which may be enough like our Sun to give their planets an environment like that of our own Solar system—then how many of these planets may be enough like Earth to develop our kind of life—then how many of the life-bearing planets may develop civilizations and so on. One of the most assiduous and persuasive of these calculators is the American astronomer Carl Sagan, who is perhaps the outstanding "exobiologist" (one who studies the possible nature of extraterrestrial life) in the world.

The American astronomer Stephen H. Dole, in his book *Habitable Planets for Man,* published in 1963, took into account all the factors of planetary systems that could be deduced (sometimes shakily) from the one system we know, our own. He ended by suggesting that our galaxy might have as many as 640,000,000 Earth-like, life-bearing planets and, presumably, that there may be as many as that in each of the other galaxies, on the average.

He also concludes that there is a 50 percent chance of finding an Earth-like life-bearing planet within 22 light-years of Earth.

To be sure, since 1963 added knowledge concerning the Universe has made it seem, increasingly, to be a desperately violent place. There are quasars that are far smaller than galaxies yet far brighter, neutron stars that have arisen out of gigantic stellar explosions, black holes that inexorably swallow up all neighboring matter and give back nothing at all.

It may be that the nucleus of every galaxy is the scene of violent events that produce an environment inimical to life. It may be that we might expect life to exist only in the sparsely distributed stars in the quiet suburbia of a galaxy's spiral arms.

Only 10 percent of the stars in a galaxy are located in its spiral arms, so that the chance of life in the Universe

133

might be viewed as falling dramatically to only 10 percent of what might have been thought a little over a decade ago. Yet even so, Dole's calculations might yield 64,000,-000 Earth-like life-bearing planets in the spiral arms alone and that is not exactly a small number. And since our own Solar system is located in one of the spiral arms of our Galaxy, our near neighbors within, say, 10,000 light-years would remain unaffected by this view.

The sort of astronomical thinking that, since World War II, has made it seem that both planets and life are overwhelmingly common in the Universe has, of course, had its effect on popular thinking. Quite aside from science fiction stories which have very frequently depended on plenty of planets and a great deal of life to add thickening to their plots, new cults involving extraterrestrial life have arisen.

There is, for instance, the flying saucer cult, which grew up immediately after World War II. This assumes that spaceships from other worlds are regularly observing Earth. As it becomes less and less possible for anyone, even cultists, to suppose that there are space travelers from any of the worlds in the Solar system, there has been a tendency to assume they come from worlds circling other stars.

In the early 1970s came the cult centering on Erich von Däniken's *Chariots of the Gods.* This supposes that space travelers from outer space, landing on Earth in prehistoric times, taught mankind its technologies, are responsible for some of the artifacts of primitive cultures, and may even have contributed alien genes to Earth's native life.

Scientists do not take these cults in the least seriously, but neither do they entirely discount the possibility of communication with extraterrestrial life. The difficulties in the way are, however, perhaps too easily dismissed by the cultists.

First, the distances between stars remain a strongly insulating phenomenon. Even if we suppose with Dole that there is a 50 percent chance of finding an Earth-like life-bearing planet within 22 light-years, that is not exactly close. It may be a tiny distance compared to the size of the Universe, or even the size of the Galaxy, but on the other hand it is 32,000 times the distance from here to Pluto our system's farthest planet.

To be sure, we might conceive of ways of getting round

the barrier of the speed of light as maximum. There is no evidence at all, however, that any of these conceivable schemes for faster-than-light travel exist, or can exist. For the moment, then, we must assume that the speed of light remains the limit.

Even so, if we are willing to take enough time to begin encroaching upon eternity, we may attempt to communicate with extraterrestrial life, nevertheless. On March 2, 1972, the Jupiter probe Pioneer 10 left Earth. It passed Jupiter on December 3, and then moved beyond and, in 1984, will leave the Solar system altogether.

With it, it carries a 6-by-9 gold-covered aluminum slab designed by Sagan and his fellow astronomer Frank Donald Drake, and drawn by Mrs. Linda Sagan. The slab carries information giving the nature of the Solar system from which it originated and of the living beings who originated it, but it may travel millions of years before it is picked up.—And it may *never* be picked up.

To send a material object is, however, an unlikely method of attack on the problem of communicating with extraterrestial life. When astronomers think of communicating with other planetary systems, they usually think in terms of massless signals such as light or radio waves which travel at the speed of light, thousands of times faster than Pioneer 10.

Thus, in the early 1960s, Drake supervised a brief attempt to detect radio signals from the direction of several stars that were thought to be Sun-like enough to have some chance of possessing life-bearing planets, and close enough to have some chance of allowing radio signals to reach us in detectable intensities.

Nothing was detected, but the thought remains. In 1969, when the English astronomer Anthony Hewish picked up for the first time very rapid pulses of radio waves of a type like nothing ever detected before, the first exciting thought was that it was a message of intelligent origin. It was referred to at first as an "LGM" phenomenon, the initials standing for "little green men."

It turned out, however, that the signals were received from rapidly rotating neutron stars—fascinating in themselves but not nearly so fascinating as the discovery of extraterrestrial life would have been.

Another uncertainty remains, and perhaps the most serious. Even though planets may be a universal accompaniment of stars, and even though life may be a universal accompaniment of Earth-like planets, how sure can we be that the development of intelligence in the course of evolution is inevitable?

On Earth, life existed for some three billion years before any single species developed a large brain in an environment that made the development of a technological civilization possible.

But was such a development inevitable? Might not life on Earth have continued indefinitely without developing intelligence and a civilization? Is it possible that intelligence, unlike life itself, is the product of an excessively rare combination of events?

And even if intelligence will develop frequently on different worlds might it not be that once it evolves, then (judging from our own sad example) it may, in quick order, inevitably destroy itself?

If this were so, then, for all we can at present say, the dream of communication with life forms in other planetary systems may be hopeless. It may be that out there are only civilizations that have not yet come to be, or that have briefly come and quickly ceased to be; and that we can talk only to ourselves during the brief interval before we ourselves, as a technological civilization, cease to be.

And yet the sanguine Sagan speculates in reverse, and wonders whether there might not be a widespread Empire of the Stars in which intelligent beings communicate and interact by way of scientific techniques undreamed of by us.

He speculates, for instance, that advanced science may make it possible to use black holes to link together so space-vaulting a civilization; that devices may be used to plumb the depths of black holes where the laws of nature (as we know them) lose their validity, and where instantaneous travel across vast stretches of space, and even forward and backward in time, may be possible.

This sounds rather less plausible, if anything, than flying saucers and ancient spacemen do, but what if Sagan's dreams are true?

Imagine, then, how incredibly exciting it would be to touch, at last, even the outermost fringes of this god-like knowledge of life beyond.

15 THE BEGINNING AND THE END

FOREWORD

In 1972, the publishing house that puts out Psychology Today *was planning to put out a new college text in psychology by the same title. It occurred to the editors to have me write short 500-word essays introducing each of the more than thirty chapters; essays that, in each case, had something to do with the subject matter of that chapter. I accepted the task cheerfully—too cheerfully. Every once in a while, the silly pride that induces me to accept any task that doesn't strike me as an outright impossibility gets me into trouble. I had a terrible time thinking up the essays and every time I got a new batch of chapters I would go around in a long-faced depression.*

I managed, though, completing the last two or three on schedule while I was in the hospital for my one and, so far, only operation.

As I kept sending in the essays, however, I suppose the editors, unaware of my agony and judging only the results, decided I wasn't working hard enough and gave me another task. They wanted an imaginative essay on the beginning and the end of the Universe for another college text, to be called Physical Science Today. *It was supposed to be an area of relaxation in an otherwise unbroken stretch of hard ratiocination.*

Since it was relaxation they wanted, I had no trouble. What else is there to do with the cheerful sound of tapping keys in your ears but relax?

We stand here on the surface of the Earth and look out at the heavens. The nearest object to us is the Moon, which we see without trouble. It is at an average distance from us of 237,000 miles and, after some six thousand years of what we call civilization, we have managed to reach it.

The farthest known object is a quasar thought to be about five billion trillion times as far from us as the Moon is. It can be seen only in a high-powered telescope by someone who knows where to look. We are probably quite safe in saying man will never reach it.

Time passes for us here on the surface of the Earth and few of us have any reasonable chance of living a hundred years. But the Earth has already been in existence nearly fifty million times the length of an extreme lifetime, and the rest of the universe has been in existence for much longer still.

With our consciousness trapped in a point of space and at a moment of time, we have nevertheless managed to evolve, to our own satisfaction, a plausible picture of a universe enormously large and enormously durable.

But curiosity never rests. It pushes always at the boundaries of the known. If the universe is enormous, where does it end, if it ends at all? And when and how will it end, if it ends at all? And when and how did it begin, if it began at all?

The point at which we begin in trying to determine the nature of the universe in space and time is at one observed fact. The spectral lines of the distant galaxies are all displaced toward the red. Indeed, the dimmer the galaxies and, therefore, the farther away (we suspect) they are, the greater this "red shift."

The most logical way of interpreting the red shift is to suppose it to be evidence of the recession of the light source. The distant galaxies, in other words, are all receding from us, all moving away from us; and the farther away from us they are to begin with, the more rapidly they are receding from us and moving even farther away.

To put it more simply, the universe seems to be expanding.

In that case, the simplest speculation about the future that we can advance is that the universe will continue to expand and that the galaxies will continue to move farther and farther apart from each other. Of course, we must modify this view by realizing that galaxies tend to exist in clusters and that within the clusters gravitational attraction predominates over the tendency toward expansion.

But the clusters separate with time and the average distance between them grows ever more enormous. The separations would be so large eventually, if this expansion continued indefinitely, that no foreseeable trick of instrumentation would enable one cluster to be visible to the questing minds located on some planet within another cluster.

This is the kind of end we can expect for an "indefinitely expanding universe." It is not an *absolute* end, for the galactic clusters would continue to exist and continue to drift away from each other for perhaps all eternity. However, when a state is reached such that there is no further perceptible change to some observer, we can call it an "effective end." In this essay, I will be dealing chiefly with effective ends, rather than absolute ones.

At the effective end of an indefinitely expanding universe, we would be surrounded by nothing detectable but the twenty or so galaxies of our own local cluster. All beyond, as far as we could see with any instruments within reason, would be nothingness, except for intergalactic gas and dust. In Figure 1, I present a simple diagrammatic expression of the picture posed by indefinite expansion.

But will we see even the galaxies of our own cluster? Will they remain essentially unchanged during the course of the slow billions of years during which all the other galactic clusters will drift away from us beyond the far horizon?

Well, everything we see, everything we do, is a manifestation of energy transfer, and the "first law of thermodynamics," which is believed to be valid through all of space and through all of time, says that the total energy of the universe is constant. (This is also referred to as the "law of conservation of energy.")

If the total energy of the universe must remain constant,

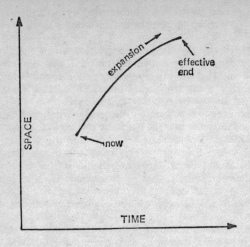

FIGURE 1
THE INDEFINITELY EXPANDING UNIVERSE (FUTURE)

then may it be that the galaxies will always shine, always live? Ah, but the first law also states that energy, while it can be neither created nor destroyed, can change its form; and the "second law of thermodynamics" (also believed to be valid through all of space and time) dictates something about the nature of the change. The second law states that less and less energy can be converted to work, as time progresses. Another way of putting it is that a quantity called "entropy" (which represents evened-out energy that can't be made use of) steadily increases with time, tending always toward a maximum.

What this means is that the universe, and every part of it, is running down. We on Earth mask this fact because we can always rebuild. When water runs downhill, we can pump it uphill again. We can reverse the process, however, only at the expense of other energy sources that have not yet run down—chiefly the energy of Solar radiation. But the Sun keeps shining only through the continuing consumption of its hydrogen. That hydrogen supply will someday give out. The Sun will expand to a red giant and then collapse to a small and very dense "white dwarf."

All stars will eventually collapse into white dwarfs, or into still smaller and denser "neutron stars," or into even

140

smaller and denser "black holes." In the end, then, perhaps before all the other galactic clusters have drifted completely away or perhaps afterward, all the stars of all the galaxies will have dwindled into dense pygmies. Only collapsed stars and thin interstellar gas would exist—and exist—and exist—and that would be the effective end of an indefinitely expanding universe in which entropy is indefinitely increasing. And from the Earth (if we imagine ourselves and it still to exist) we could see nothing at all except perhaps for a tiny red glow from the dot of light that was once our Sun.

And if that is the far future, what about the far past? Suppose we run the film backward, so to speak, and look into the past; suppose we carry the curve of Figure 1 leftward, beyond the now point in the other direction.

Moving forward in time, we have an expanding universe; moving backward in time, we would have a contracting universe. We would imagine, as we probe deeper and deeper into the past, that the galaxies are coming ever closer, that entropy is steadily decreasing toward a minimum.

If we go backward in time far enough, all the matter of the universe can be imagined as having contracted to the point of being collected into a single vast, dense ball, the "cosmic egg." If we place ourselves, in imagination at the point in time where the cosmic egg exists. we can think of it undergoing a huge cataclysmic explosion. It is this "big bang" which would have started the universe on its career of expansion, an expansion that still continues today.

The cosmic egg would represent an "effective beginning" for the universe. It might conceivably have existed for eons before the explosion, for all eternity perhaps. but it is at the time of the explosion that we can imagine changes beginning to take place, and it is the onset of change that offers us an effective beginning (see Figure 2).

If we picture the universe now as beginning at the cosmic egg, with minimum entropy, and ending at a scattering of neutron stars, with maximum entropy, it would seem that the beginning is the less satisfactory of the two extremes.

After all, how did the cosmic egg come into existence? How did matter get to be squeezed so tightly together in the first place? Surely, a cosmic egg is a highly unstable arrangement, as is shown by the fact that everyone who

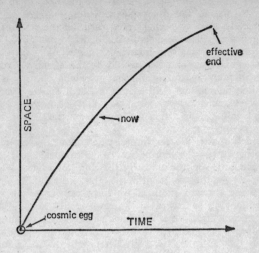

FIGURE 2
THE INDEFINITELY EXPANDING UNIVERSE (PAST AND FUTURE)

speculates about it invariably deals with its vast explosion. If it is so unstable, why should it exist in the first place? If, on the other hand, it existed because it *was* stable, why then did it become, at some particular moment in time, so unstable as to explode?

These are difficulties that have led some astronomers to try to picture a universe in which the cosmic egg never existed at all. They cannot deny that the universe is expanding and that the galactic clusters are continually separating. They suggest, though, that as the clusters separate, new matter forms in the universe and collects together, making up new galactic clusters between the separating old ones.

If we imagine ourselves watching such a universe from some point within it, old galaxies are forever drifting away beyond the horizon, but new ones are always forming in the foreground. The fine details of what we see, the individual galaxies and their shapes and exact positions, may change, but the overall view does not.

If we imagine ourselves watching the same universe in which time is running backward, the distant galaxies approach but some in the foreground are always unforming and vanishing as older ones come into view from beyond the horizon. Again the fine details change, but the overall view does not.

In such a "continuous-creation universe" or "steady-state universe," size is infinite and duration is eternal, and the part we can see never changes. There is no beginning and no end, or if we consider lack of change to mark an effective beginning and an effective end, the steady-state universe is all beginning and end and nothing else (see Figure 3).

At the present moment, however, we must eliminate the steady-state universe as a possibility. There is some observational evidence we can draw on (the radio-wave background that exists in every direction, for instance) that seems to make it reasonably sure that there was indeed a big bang in the past and that the cosmic egg did at one time exist even if only momentarily. Which means we are stuck with its paradoxes.

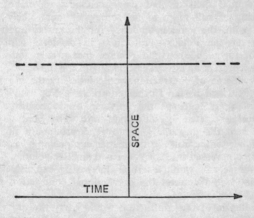

FIGURE 3
THE STEADY-STATE UNIVERSE

Suppose we tackled the problem from the other direction. Suppose we ask ourselves what sort of a beginning of the universe would be a stable one. What would a universe be like before any of its present organization existed? What form could we accept as both ultra-primitive and ultra-stable, so that we could imagine that form as having

143

existed for an indefinite period before changing into our universe of today?

One possibility, perhaps the easiest to envisage, is that the universe existed, to begin with, as an exceedingly thin gas, made up of the simplest possible atoms—those of hydrogen. This is, in essence, the form in which we find the universe existing in the space between the clusters of galaxies. We are supposing, then, that the universe began as nothing but intergalactic space, which would certainly be stable over enormous periods of time.

Such an exceedingly thin gas would, however, be subject to its own vastly diffuse gravitational field. Random motions of the atoms would be a little more likely to bring them together than to separate them, since gravitational influences would be acting in favor of the first alternative and against the second. Once they were together, interatomic forces might tend to keep them together.

Any clustering of atoms would serve to intensify the gravitational field in that neighborhood and make the gathering of still more atoms somewhat more likely. At some point, the general tendency for the gas to contract (whatever the details—uniformly, in clusters, with or without turbulence) would become measurable and that would represent an effective beginning of the universe.

Any degree of contraction would intensify the gravitational field and hasten further contraction. For many billions of years, perhaps the universe would continue to contract at a steadily increasing rate. The contraction would heat the universe generally and produce a higher and higher temperature in the matter composing it, as that matter compressed into a smaller and smaller volume. The temperature rise would exert an increasing expansive effect that would counter the gravitational contraction and begin to slow it down.

The inertia of the contraction would carry the universe past the point where the temperature effect would just balance the gravitational in-pull, if the universe were not contracting. The universal contraction would continue to some minimum volume, represented by the enormously dense cosmic egg, to the point where the expansive effect of temperature would finally gain control. The substance of the universe would then be pushed outward, faster and faster, until the accelerating force becomes the big bang.

The general picture is rather like that of an elastic ball dropping from a height onto a hard surface, compressing

144

itself, and then, having the force of the compression reversed, hurling itself up again in a high-flying bounce.

In this view, the universe begins at a point of general emptiness, with matter distributed as single hydrogen atoms. It also ends at a point of general emptiness with matter distributed in dense clots as collapsed stars. The cosmic egg is the unstable and momentary midpoint. This is a "hyperbolic universe" (see Figure 4).

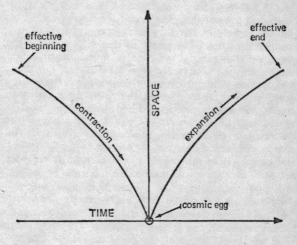

FIGURE 4
THE HYPERBOLIC UNIVERSE

But now it is the ending that raises a question. The beginning with its thin scattering of hydrogen atoms is very simple, but the end with its thin scattering of collapsed stars seems less simple. In the end there are regions of extremely intense gravitational fields. Can they, growing large enough, reverse the expansive effect? Expansion overcame contraction at the time of the cosmic egg; is it possible that contraction will overcome expansion at the time of the collapsed-star scattering?

If so, is our present state of universal expansion to be succeeded by another contraction, somewhat similar to that which preceded the formation of the cosmic egg? And can we envisage the eventual formation of another cosmic egg?

The picture is a troublesome one. The collapsed-star universe is close to entropy maximum, while the cosmic egg is close to entropy minimum. The second law of thermodynamics says that entropy must increase with time; events must go from cosmic egg to neutron star.

Ah, but the second law has been observed and tested only within the context of an expanding universe and we have no right to insist on its applicability to a contracting one.

As the universe contracts, the radiation emitted by the galaxies, as viewed from the center, undergoes a shift toward the violet and therefore gains in energy in the direction of contraction. The extent of that "violet shift" (the opposite of the galactic red shift so prominent in an expanding universe) increases as the velocity of approach grows greater with accelerating contraction. The energy poured into the center of the universe by the intensifying gravitational field is compressed and itself intensified. Under the lash of that energy, the downhill changes that take place in an expanding universe are reversed. The universe winds itself up again.

Of course, once the cosmic egg is formed, it explodes once more and we would go through the cycle again, and again, and again. Instead of a hyperbolic universe, once-in-and-once-out, we would have an "oscillating universe" or a "pulsating universe," one which expands, contracts, expands, contracts. and so on indefinitely.

But will each period of the cycle be exactly like the one before? Can it be, for instance, that in the process of expansion, a great deal of matter is turned into energy within the various stars? This energy, in the form of photons, neutrinos, gravitons, and so on, might be viewed as escaping and streaking outward forever. The matter that ends up as neutron stars at the point where contraction begins again, may therefore be considerably less than the mass of the original cosmic egg.

When the cosmic egg forms again, then, it will be smaller than the one before and will explode less violently. The next period will then be shorter and result in a still smaller cosmic egg, and so on indefinitely.

Finally, a cosmic egg will form that will be so small it will not store enough energy, in forming, to bring about an explosion once more. It may just pulsate in a gradually dying tremor, itself no larger than a single moderately large neutron star, and that would be the effective end of such a "damped-oscillating universe."

And what about the beginning of such a universe? If, in our minds, we run the evolution backward, we find that the cosmic egg formed at each compression state of the cycle is larger than the one before and the period between successive cosmic eggs longer. Eventually we would imagine a cosmic egg so huge and an explosion so vast that gravitation never manages to overcome the outward thrust. There would be a *first* compression.

Going back to the effective beginning of a damped-oscillating universe, then, we would have a thin distribution of matter, not in the form of neutron stars (which might imply a previous cosmic egg), but in the form of hydrogen gas. In other words, the damped-oscillating universe could be pictured as beginning in the same way the hyperbolic universe begins, but instead of forming one cosmic egg, forming a whole series of ever-smaller ones; many bounces in place of one (see Figure 5).

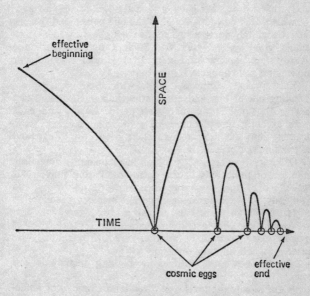

FIGURE 5
THE DAMPED-OSCILLATING UNIVERSE

Yet such a damped-oscillating universe doesn't look entirely reasonable in the context of Einstein's theory of relativity. According to relativistic notions, space is curved by an amount depending on the nature of the general gravitational field of the universe. Photons and neutrinos escaping from the stars and traveling in "straight lines" are not really traveling in straight lines in the Euclidean sense. They are not streaking outward forever, never to return. They are, instead, following a very gently curved path, the curvature of which is, in fact, what we mean when we say that "the universe is curved." In a sense, radiation does not leave the universe, and is not lost to it.

In the contracting phase of a universe, the gravitational field intensifies and the curvature of space generally grows more marked. As the galaxies approach one another, the photons and neutrinos spiral inward, too, following sharper and sharper curves.

The result may well be that there is *no* loss of matter in the course of a cycle, and that each cosmic egg is just as large as the one before.

If that be so, we have a "steady-oscillating universe" in which each period is equal to the one before and the one after, and the number of periods can be viewed as infinitely great (see Figure 6). Such a universe, like the steady-state universe of Figure 3, is without beginning or end, but is not changeless, merely periodic.

Each period begins and ends with a cosmic egg, expanding out of one and contracting into one. Or you can say, with equal truth, that each period begins and ends with a thinly spread out volume of collapsed stars, contracting into the intermediate stage of a cosmic egg and expanding out of it again.

Of course, we are only justified in picturing an unending number of pulsations in the steady-oscillating universe if we think of time as proceeding onward uniformly and endlessly. If we suppose each period between cosmic eggs to last a trillion years (just to use a round number), then we would say that a trillion years after the most recent cosmic egg there will be another one and a trillion years after that, still another one, and so on.

Are we, however, justified in thinking of time in this way?

At the subatomic level, the direction of time is of no account. All the laws of particle behavior are the same whether time moves "forward" or "backward." This means

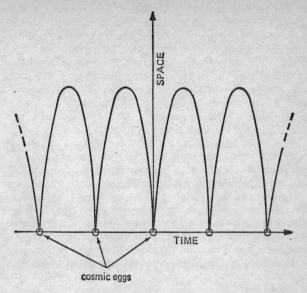

FIGURE 6
THE STEADY-OSCILLATING UNIVERSE

that if we were a subatomic particle living among other
subatomic particles, we would have no way of telling
whether time was moving forward or backward. We would
merely define one direction as forward and then the other
direction would be backward.

In our own universe at the macroscopic level of organization, it is convenient to consider time as moving in one
direction only, because entropy is steadily increasing so
that all changes seem to be of one general kind. Objects
suspended in midair always drop; objects hotter than their
surroundings always cool off; and so on. We arbitrarily
call this apparent motion of time "forward."

If we run a movie film backward, we can see a broken
vase reassemble itself out of its pieces, and fly up from
the floor to the stand; we can see a diver rise from the
water to the diving board, while the water splash gathers
together. We see at once that this is not the way things
happen in what we call "real life." That is what would
happen if time flowed "backward."

149

But in a contracting universe, where all the galactic clusters are coming together and the universe is winding up with entropy steadily *decreasing*, the film, so to speak, *is* running backward. The events that take place are what would take place if time moved in the reverse direction.

In that case, let us imagine ourselves starting at a cosmic egg (any cosmic egg) in a steady-oscillating universe such as that shown in Figure 6. The universe would expand for half a trillion years with time moving backward. It would end at the same time as it began.

Instead of imagining an infinite number of pulsations occurring over an eternity, we might imagine a single expansion, followed by a single contraction, over a time of half a trillion years and back (see Figure 7).

To imagine the picture at its most simple, we might conceive there to be a single universe existing in four dimensions, the three spatial dimensions plus the fourth, time dimension. We might conceive it to be a "hypersolid," beginning as a small volume filled with cosmic egg at time-zero, and ending, cornet-fashion, as an enormous

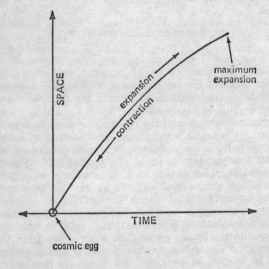

FIGURE 7
THE STEADY-OSCILLATING UNIVERSE (BIDIRECTIONAL TIME)

volume containing a thin scattering of neutron stars at time-half-a-trillion-years.

Everything in such a fixed universe would be, in the larger sense, permanent. Somehow, though, our consciousness progresses steadily forward through the hypersolid and, at each instant of time, is conscious of a different three-dimensional cross section. It is this progress of consciousness forward at a fixed speed, so to speak, that gives us the illusion of change and time.

If we could then imagine a consciousness greater and less evanescent than our own, sweeping forward from cosmic egg to maximum expansion, then sweeping back to cosmic egg again, we would have the illusion of a universe endlessly pulsating but always repeating itself at every stage exactly.

And yet the notion of a fixed universe does not quite fit the very fundamental natural law known as the "Heisenberg uncertainty principle." This limits the precision with which any measurement can be made or any effect predicted. If we consider the state of the universe at any instant of time, there are different states that are equally probable at some other instant of time in the future.

As the consciousness we are imagining sweeps backward through time and then forward again, it may this second time pass through a somewhat different universe of equal probability. In fact, though it sweeps back and forth an infinite number of times, it may in each sweep never quite repeat the details of the previous sweep, or any previous sweep in that direction.

We might imagine that in the time interval between time-zero and time-half-a-trillion-years, there lie an unending number of universes, each different in detail from all the rest and separated by something more subtle than time and space. In this way, we substitute for a steady oscillating universe unending in time, a steady-oscillating universe unending in probability states.

But in either case, whether the universe oscillates equally and unendingly in time or in probability states, what is unending is the universe's material content. Whether the matter is compressed into a cosmic egg, or thinned out into separate neutron stars, it is always there. We are surely entitled to pass on in our questioning from the matter of the beginning and end of the organization of the universe (which is what we've really been doing so far) to the beginning and end of the matter that makes it up.

Actually, this is not so difficult a problem as it might sound. Matter is *not* eternal. We can make it begin and end in the laboratory.

Almost every particle has its "antiparticle." A particular particle and its antiparticle are identical in all respects except for electric charge or magnetic field or both. Thus, a proton carries a positive electric charge of a certain fixed size, while an antiproton carries a negative electric charge of the same fixed size and differs from a proton in no other respect, except that the direction of its magnetic field is also reversed.

Particles can combine to form nuclei, atoms, molecules —and the corresponding antiparticles can do the same. Ordinary particles can combine to form ordinary matter; antiparticles can combine to form antimatter.

If ordinary matter is considered as existing by itself, it cannot be entirely destroyed. A small fraction of it (about 1 percent) can be converted into energy if the particles are particularly well packed, and it is this energy formation through better packing that is the source of the radiation of the Sun and of stars like it.

The same would be true of antimatter if that were considered as existing by itself. In an antiuniverse, antistars would also radiate energy produced through better packing.

Particles of matter can, however, combine with those antiparticles of antimatter analogous to themselves to produce "mutual annihilation." All the matter in both sets of particles would no longer exist—as matter. They would be converted to energy in the form of energetic photons. (Photons, the constituent particles of radiation such as visbile light, have no antiparticles. They are equally at home in both the universe and the antiuniverse.)

The reverse is also true. Energetic photons can be converted to matter—but never to matter alone, and never to antimatter alone. A particular quantity of energy will form particles of matter of a particular type in particular numbers, but will inevitably also form the same number of the equivalent antiparticles of antimatter.

If the matter of the universe is not to be considered as having been eternally in existence, then, the logical alternative is to suppose it to have been formed out of energy.

152

But if matter was formed out of energy, an equivalent amount of antimatter must also have been formed.

And in that case, where is the antimatter? Our planet is composed entirely of matter, with no significant quantities of antimatter at all. The Moon is matter, too; the entire Solar system is. In fact, as nearly as we can tell from the cosmic-ray particles that reach us from outer space, our entire Milky Way galaxy is matter—perhaps even our entire universe.

Where is the antimatter then?

It is not likely that after the formation of particles and antiparticles, these would remain well mixed—for then they would interact after a comparatively short interval of time and undergo mutual annihilation back into energy. If matter and antimatter were to be expected to form out of those particles and antiparticles, and if they were to be expected to continue to exist, they would have to be separated from the moment of formation, and they would have to continue to remain separated by distances great enough to avoid significant amounts of interaction over extended periods.

If the matter within our galaxy were partly antimatter we would probably be aware of radiation from various directions arising from the inevitable occasional mutual annihilations that would be taking place. We can be reasonably sure that the galaxy is just about entirely matter, if only because such radiation is *not* detected. It may also be that any galaxy, or even any cluster of galaxies, is entirely matter—or entirely antimatter.

It may be, however, that the universe consists of clusters of galaxies and anticlusters of antigalaxies.

We can't be sure which are which—except that our own is matter by definition. The photons and gravitons galaxies emit would be the same whether those galaxies were composed of matter or of antimatter. To be sure, galaxies would emit neutrons while antigalaxies would emit antineutrons, but both neutrons and antineutrons are extremely difficult to detect. If two galaxies were in contact or near-contact and the pair gave rise to unusual quantities of radiation, we could conclude that one was a galaxy and one an antigalaxy, but not be able to tell which was which.

But then, it may be that all the galaxies of the universe are matter, and that a similar enormous collection of antigalaxies exists somewhere far beyond the range of detection by our instruments and forms a separate "anti-universe."

How could the separation of matter and antimatter into galaxies and antigalaxies, or into a universe and an anti-universe, have come about?

For one thing, charged particles and antiparticles react in opposite fashion to a given electromagnetic field. If in the process of explosion, the cosmic egg produced a huge electromagnetic field, all charged particles would swerve off in one direction while the corresponding antiparticles would swerve off in another. Then, too, there is a gravitational attraction between all particles of matter and pre-sumably, between all antiparticles of antimatter. This might mean there is gravitational repulsion between matter and antimatter, though, to be sure, no such effect has yet been detected in the laboratory. (Nor has such an effect been ruled out in the laboratory, either.)

It is useless to try to work out the details; whether the separation was worked out by electromagnetic forces, gravitational ones, or both; whether it took place in one grand sweep, or little by little with numerous interactions and annihilations en route; whether it resulted in galaxies and antigalaxies or a universe and an antiuniverse; whether the separation took place in the course of seconds or of millions of years.

The fact remains that the only way we can avoid the assumption that the matter of the universe has always existed is to suppose it arose out of energy and in that case there is antimatter somewhere, too, and the neatest solution, however it came about, is to suppose a universe and an antiuniverse.

If we do this, we can perhaps avoid a second difficulty I have so far skirted over.

If we imagine a contracting universe, made up of matter alone, collapsing into a cosmic egg, we may have trouble envisaging anything to stop the collapse. Individual stars can collapse into white dwarfs, say, ten thousand miles across; or beyond that into neutron stars, say, ten miles across; or beyond that into black holes in which the volume can shrink to zero while the density and gravitational intensity rise to indefinitely large figures.

The larger the shrinking star, the more likely it is to plunge past the intermediate stages and into the black hole ultimate. The black hole has a gravitational field so intense that nothing can ever get out again, not even massless particles such as those of light and other radiation. We know of no mechanism that would make a black hole explode.

It seems quite likely that a star, large but not enormously large, contains enough matter to make black hole formation possible. Certainly the entire universe contains enough matter for the purpose. Well, then, when the universe contracts to form a cosmic egg, what keeps it from plunging past the cosmic egg to form the hugest black hole that can be conceived and ending all and everything in that fashion. The expansive properties of ordinary matter, compressing, would not seem enough to overcome the potentiality of black hole formation.

But suppose instead that as the universe contracts, galaxies and antigalaxies are forced closer and closer together. Or suppose that as the universe contracts, the antiuniverse contracts also, and as they travel back through time, they come closer and closer together. In either case, as contraction proceeds, the incidence of interaction and mutual annihilation increases. The mutual annihilation at a steadily increasing rate not only produces far more radiation and therefore expansive tendency than could be produced by any contraction involving matter alone (or antimatter alone) but also destroys mass and therefore lowers the gravitational intensity that powers the contraction.

As a result, the contraction may well stop short of black hole formation and produce instead a kind of cosmic egg that is pure energy and not either matter, antimatter, or a mixture of the two. Now we have a steady-oscillating universe beginning as an energy-egg and producing two universes (see Figure 8).

(Remember that conditions in the antiuniverse would be no different in essentials from those in the universe. An antiman in an antiuniverse would be perfectly justified in defining his own as a universe and calling ours an antiuniverse.)

Does the double-universe of Figure 8 now answer all questions? Unfortunately, it doesn't. We can still ask ourselves—

Where does all the energy, out of which the universe and the antiuniverse have arisen, come from? Must we view that energy as constant and eternal, without beginning or end?

To be sure, this is exactly what the law of conservation of energy asks us to do, but how far can we trust that

155

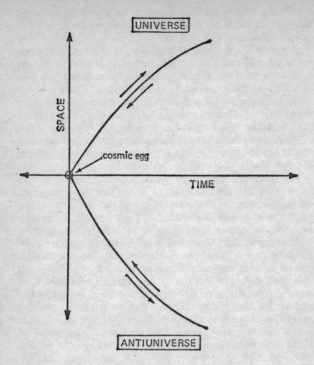

FIGURE 8
THE STEADY-OSCILLATING DOUBLE-UNIVERSE

law? To what extent must we consider it a generalization applying only to the local conditions of our own universe in its present state?

To see what I mean, suppose we consider a law similar to it; the law of conservation of momentum (where the momentum of any object is the product of its mass and velocity). By the law of conservation of momentum, momentum, like energy, can neither be created nor destroyed.

And yet—a bullet at rest in a rifle has zero momentum with respect to the Earth, while the same bullet in flight has a great deal of momentum. Has not the pulling of the trigger succeeded in creating momentum out of nothing?

156

No! For while the bullet speeds in one direction, the rifle has, through recoil, moved in the opposite direction. If we define momentum in one direction as positive and in the other as negative, then we will have $+x$ momentum for the bullet and $-x$ for the rifle. The two will add up to zero, so that the zero momentum before the firing will equal the zero momentum after the firing, if, in both cases, rifle and bullet are considered together. The whole "system" can neither gain nor lose momentum, though individual parts of a system can.

The same is true of "angular momentum." Something that is not rotating will begin to rotate in one direction if another part of the system undergoes an equivalent rotation in the opposite sense.

There is also such a thing as a law of conservation of electric charge. A proton with its positive electric charge cannot be created out of energy (which has no charge) unless an antiproton, with a negative electric charge, is also created. The two charges, positive and negative, are equal in size and add up to zero, so that no *net* electric charge has been created.

Can this sort of thing be applied to energy, too? Can there be "positive energy" and "negative energy" and can equal quantities of both be created out of "zero energy," just as equal quantities of positive and negative momentum can be created out of zero momentum and as equal quantities of positive and negative electric charge can be created out of zero electric charge.

It must be admitted that there is nothing anywhere in theory or in observation to support the possibility of the existence of two opposite kinds of energy, and yet suppose these exist. What might the consequences be?

Let's begin by remembering that all the particles we know of as making up the universe represent some form of energy. In that sense, everything we can detect in the universe, without exception, is energy in one form or another arranged in some form of interrelationship. If the universe contained no energy at all there would be nothing left (as far as we know) that we could detect. Nor would we exist or anything that (as far as we know) could serve as detector.

We can express this by saying that positive energy and negative energy would interact to produce *nothing!*

But if that is so, then the reverse ought also to be true. A quantity of nothing might suddenly become equal

157

quantities of positive energy and negative energy. The two states of existence—nothing on one side; equivalent amounts of positive and negative energy on the other—might have equal probabilities of existence. For that reason, nothing could give rise to double-energy on a purely random basis.

In fact, if we imagine a field of nothingness, random effects might be continually producing bubbles of negative and positive energy, always in equivalent amounts. Some bubbles might be larger than others and it would seem reasonable to suppose that the larger the bubbles, the less frequently they form.

When the bubble-pair is formed, it may be that there is some tendency for the two to separate—a mutual repulsion. The larger the bubble-pair, the farther they move apart, and the longer it takes them to come back together and interact—turning back into the nothing from which they came.

Suppose that the quantity of positive energy formed is roughly equal to all the energy present in the universe and antiuniverse I represented in Figure 8. There would then be enough negative energy produced to form a negative universe and a negative antiuniverse of equal size.

What's more, it isn't hard to conceive that the bubble of positive energy formed is the equivalent of the energy-egg I mentioned earlier (a cosmic egg made up of energy only). It explodes to form the universe and antiuniverse, which expand and then contract again in a period of, say, a trillion years.

Meanwhile the bubble of negative energy formed is an antienergy-egg. It explodes to form the negative universe and the negative antiuniverse, which expand and then contract again in the same period of, say, a trillion years.

When the contraction stage is completed the energy-egg and the antienergy-egg are formed again. Perhaps they explode again, and after another expansion-contraction cycle, form and explode. They may do so an indefinite number of times. Or else, after a few expansions and contractions, or even after only one, the energy-egg and the antienergy-egg interact and sink to nothingness again.

In what way, though, are the positive and negative universes different and what keeps them separate? Here we are in the realm of fantasy alone, for we have nothing to guide us.

We can reason, though, that the positive and negative

universes must be identical except in some key respect in which they are opposites, and that these opposites must have no internal effect. Just as an antiuniverse would seem perfectly normal to an antiman; a negative universe would seem perfectly normal to a negative man; and a negative antiuniverse would seem perfectly normal to a negative antiman.

In antimatter, the electric charge on a particular anti-particle is the opposite of that on the analogous particle. In negative matter, it might be hypothesized, the direction of time is reversed, as compared with that in positive matter.

In the universe and antiuniverse alike, time goes in that direction we arbitrarily call "forward" during the period of expansion and "backward" during the period of contraction. In the negative universe and negative antiuniverse alike, time would go "backward" during the period of expansion and "forward" during the period of contraction.

This difference in time direction could not be detected from within a universe where the difference affects all things alike, including the detector.

It is only when universe and negative universe are compared from outside by an observer who is a member of neither that an effect would be detected. That effect would be this: that the time interval between the two is continually increasing as long as both are in the contracting stage.

Just as the universe is separated from the antiuniverse (and the negative universe from the negative antiuniverse) in space, so the two positive universes are separated from the two negative universes in time. The result is the "quadruple-universe" shown diagrammatically in Figure 9.

In such a case, we can finally imagine an absolute beginning and end to all aspects of the universe. There is an absolute beginning when the quadruple-universe is created out of nothing; and there is an absolute end when the quadruple-universe sinks back into nothing. In between, the quadruple-universe may oscillate once, twice, or an indefinite number of times.

Once the quadruple-universe sinks back into nothingness, it, or another like it, may begin again. In fact, it would be logical to suppose that it must begin again.

Random factors must be producing bubble-pairs of energy here and there in nothingness continually.

Indeed there may be numerous such bubble-pairs existing. If the nothingness is unending (and why should it not be?) there can well be an infinite number of quadruple-universes existing along with ours and separated from ours by something more subtle than time or space.

And yet just as I feel that I have evolved a scheme in which all questions are answered, in which there is an absolute beginning and an absolute end, I find that I can question still further—

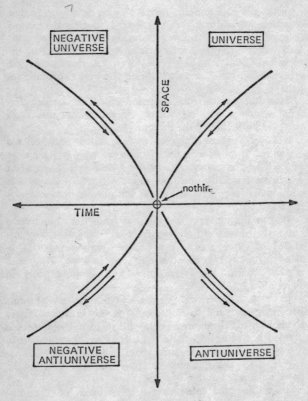

FIGURE 9
THE STEADY-OSCILLATING QUADRUPLE-UNIVERSE

What is this nothingness out of which all the quadruple-universes arise and into which all sink?

Is it really nothingneess or is it something other than nothing which we confuse with true nothingness only because we don't know how to detect or study it? Is it something more subtle than energy, but just as real and just as other-than-nothing?

And in that case, can it exist in opposite varieties so that the quadruple-universe is really an octuple-universe sinking back to a "nothingness" that is really a still more subtle other-than-nothing?

Can we allow ourselves to wonder whether everything that exists is one of an infinite number of such objects that arises out of something more fundamental than itself, which is, in its turn, one of an infinite number of such objects that arises out of something still more fundamental, which is itself one of an infinite number of such objects that—

And so on, and so on, and so on—

Without beginning and without end.

AFTERWORD

As you see, I was sufficiently pleased with the above essay to name the entire book after it.

16 GRAVITATION, UNLIMITED

FOREWORD

Every once in a while science will make a sudden turning, and sometimes I don't become aware of the turning for a period of time. Invariably, it seems, I become aware of it immediately after I have published an article which doesn't take the new factor into account.

Thus, talk about black holes in space had been gradually growing louder ever since the mid-1960s, but even as late as January 1972, when I wrote "The Beginning and the End," I wasn't taking it seriously, and didn't particularly deal with black holes in the article.

Almost as soon as the article was published, however, it became quite obvious to me that one couldn't talk about the beginning and the end of the Universe without talking about black holes. I therefore eagerly seized the chance of writing an article on them when International Wildlife *asked for one.*

In the last dozen years, astronomers have been finding a whole series of new cosmic wonders undreamed of before. Each seems to be more freakish than the one before, and the most recent, the "black hole," would seem to be the ultimate—for in it lies, perhaps, both the beginning and the end of the Universe.

The explanation of that starts from the simplest facts—

There are exactly four known ways in which matter can interact. These four interactions, as far as scientists can tell so far, explain everything that happens.

Two of these interactions are nuclear and are called "strong" and "weak." The strong interaction is the cement that holds together the particles making up the atomic nuclei. The weak interaction governs the manner in which certain nuclei eject electrons.

These two nuclear interactions are short-range. Their effect fades off very quickly with distance and barely extends across the width of an atomic nucleus. They are of no direct importance in the ordinary world of ordinary events.

The remaining two interactions are long-range, for they drop off in intensity only as the square of the distance. They are the "electromagnetic" and the "gravitational" interactions.

Electromagnetic intensity depends upon the size of the electric charge on particles of matter. Gravitational intensity depends upon the size of the mass of particles of matter. Any particle that carries an electric charge also has mass, so any particle that is subject to the electromagnetic interaction is also subject to the gravitational interaction—but with what a difference!

Consider an electron, which is the smallest particle of matter that carries a negative electric charge. Its mirror image, the positron, carries a positive electric charge of the same size. The electron and the positron attract each other both electromagnetically and gravitationally—but the electromagnetic attraction is over 2,000,000,000,000,000,000,000,000,000,000,000,000 (two thousand trillion trillion trillion) times as intense as the gravitational attraction.

It is the electromagnetic interaction that keeps electrons in place near atomic nuclei; that holds atoms together and builds molecules; that keeps liquids liquid and solids solid; that governs all the workings of the molecules of our body and is therefore basic to life. The gravitational interaction plays no perceptible part in any of this.

And yet gravitational interaction, incredibly weak though it is, dominates the Universe and is its one overriding fact.

How can this be? —Well, the failure of the electromagnetic interaction to dominate rests on the fact that there are two kinds of electric charge: positive and negative. A positive charge attracts a negative charge, but two positive charges repel each other and two negative charges repel each other.

Whenever large numbers of charged particles are grouped together, the numbers of positive and negative charges are always roughly equal and they are always well mixed. The result is that all the attractions and repulsions pretty nearly cancel each other. The actual size of the overall electromagnetic interaction of any body, however large, depends on the excess of one type of charge over the other and this is always rather small.

In the case of gravitational interaction, however, the intensity depends only on mass and there is only *one* kind of mass. The greater the mass (that is, the larger and denser the body), the greater the gravitational interaction.

Thus, every particle in the Sun is the center of the tiniest imaginable quantity of gravitational interaction, but all the particles together produce a total intensity sufficient to hold the Earth in its orbit.

At astronomical distances, only the gravitational interaction really counts; only this imperceptibly weak interaction, which builds up and builds up and builds up as masses accumulate, controls the bodies making up the Universe. It governs the way in which planets circle suns, in which stars circle each other, in which whole clusters of stars cling together by the millions and even by the trillions, in which the galaxies themselves move among each other in clusters, and in which the Universe itself expands and contracts.

How strong can the gravitational interaction become? It increases as distance decreases. The Earth's gravitational force can be felt from the Moon, but if one measures it as one approaches the Earth from the Moon, it grows stronger and stronger.

The Earth's gravitational interaction could increase without limit as we continue to approach the Earth's center, but when we reach the Earth's surface, we can approach the center no more. The Earth's gravitational pull at the surface is as strong as we can experience it.

We can imagine ourselves burrowing through the Earth, downward toward the center. If we do that, though, we leave some of the Earth's mass above us. We are affected only by that portion of the mass that is nearer the center than we are. The intensity of the gravitational interaction actually decreases as we burrow and reaches *zero* when we are at the Earth's center.

But what if all the mass of the Earth is squeezed together into a smaller ball. You could then approach more closely to the center and still have all the mass beneath you—and the gravitational interaction would become stronger. The more the mass of the Earth is squeezed together, the more it is packed into a smaller and smaller ball, the stronger the gravitational interaction at the surface would be.

The Earth, however, cannot be squeezed smaller. Its atoms are in contact and resist being pushed closer together.

What about the Sun, though? The Sun is enormously more massive than the Earth is; 333,400 times as massive and, therefore, that much greater as a source of gravitational interaction. Under the pull of its own enormous gravity, the outer layers of the Sun press downward with such force as to smash the atoms near its center.

Ordinarily, the electrons in intact atoms, such as those atoms making up the Earth even at its center, remain a fixed and rather great distance from the nucleus, so that most of the atom might be considered empty space. In the inner layers of the Sun, the electrons are pushed inward and the empty space disappears in part.

Yet the Sun doesn't shrink as its atoms collapse. The center of the Sun is at millions of degrees of temperature and the expansive effect of this heat keeps the Sun puffed up despite the gravitational pull. As a result, the average density of the Sun is only a quarter as great as the average density of the Earth.

But the Sun remains hot at the center only because certain nuclear reactions are taking place. What happens when someday, billions of years hence, those reactions are all done. Then, when the nuclear reactions at the center fail, the Sun will collapse, suddenly and explosively.

When all, or nearly all, the atoms in the Sun break up under the gravitational pull inward, the Sun would no longer be its present size of 1,400,000 kilometers across. It might shrink to a size of only 14,000 kilometers across. It would have only 1/100 its present diameter and become not much larger than the Earth. There would still be heat enough to cause its surface to glow white-hot, and it would therefore be a "white dwarf."

Many stars have already reached a state of collapse and white dwarfs are common phenomena in the Universe. Because they are small, they are hard to see, so the first one was not detected until 1862. Even then, it wasn't till 1914 that the nature of the white dwarf was understood by astronomers.

A white dwarf has all the mass of an ordinary star compressed into a much smaller size. If a star shrinks to 1/100 its original diameter, the distance to the center from the surface is 1/100 as great as it was, and the gravitational pull at that surface is 100×100 or 10,000 times as great as at the surface of an ordinary star.

But what stops a white dwarf in its collapse and keeps it at a diameter of 14,000 kilometers?

The electrons, escaping from the smashed atoms, repel each other since all have negative charges. This repulsion stops the collapse of the white dwarf.

Imagine, however, a star larger than the Sun and one with a greater gravitational interaction. If it collapses, it squeezes the electrons more tightly together than the Sun would. The larger the original star, the smaller the resulting white dwarf.

If a star is more than 1.4 times as massive as the Sun, then the electron repulsions are simply not strong enough to stop the collapse. The negatively charged electrons are forced into the positively charged nuclei. The electric charges cancel and we end with "neutrons," which have no electric charge.

The neutrons collapse on each other till they are touching and there is no empty space left at all. The result is a "neutron star," which contains all the mass of a large star squeezed into a tiny ball perhaps only 14 kilometers across.

In the case of a neutron star, you can come so close to the center, and still have all the mass between you and the center, that the gravitational pull mounts up enormously. At the surface of a neutron star, the pull is perhaps 10,000,000,000 times as much as at the surface of an ordinary star.

It was not till 1969 that neutron stars were detected in space for the first time.

But what if the collapsing star is larger still? What if it is more than 2.5 times the Sun's mass. In that case, the gravitational interaction produced is so intense that even the structure of the neutron breaks down and the star collapses past the neutron-star stage.

Nothing is known that would stop the collapse once the neutrons go, so the collapse continues and continues (as far as we know) until the whole mass is squeezed into a zero-sized point, and the gravitational pull in the neighborhood of that point becomes indefinitely large.

The stronger the gravitational effect at the surface of a body, the harder it is for anything to break away from the body. Even light itself has difficulty leaving a body if the gravitational pull is too strong. It barely makes it away from a neutron star.

If a star collapses past the neutron-star stage, the gravitational pull at its surface becomes so large that nothing material can possibly escape from it. It becomes like an infinitely deep hole in space. Things can fall in but nothing can come out. Even light can't come out because the gravitational pull is too strong, so that if we could become aware of such a "hole" it would seem black. It is therefore called a "black hole."

Despite the fact that the black hole does not emit light, it can be detected in either of two ways.

For one thing, it is still a source of gravitational interaction, and the total interaction doesn't change as the star is squeezed into a black hole (only the intensity of the interaction in its immediate neighborhood). For instance, if two stars circle each other, and one collapses and becomes a black hole, the other keeps on in its previous orbit as though nothing had happened. If the black hole is more massive than the normal star, it is the normal star that does most of the moving; if the black hole is less massive, *it* does most of the moving.

In either case, if a normal star is moving in an orbital fashion with nothing visible to account for it, we might suspect a black hole in its neighborhood.

Second, matter in the neighborhood of a black hole tends to spiral inward and, in doing so, it loses gravita-

167

tional energy. Some of this energy is converted into "photons," the kind of particles that make up light and similar radiations. The photons produced under such extreme conditions are much more energetic than those which are associated with ordinary light, however. The photons given off by matter spiraling into a black hole are of the type we associate with X rays.

When a black hole is part of a two-star system, there is an enormous quantity of matter in the neighborhood which may fall in. Matter from the remaining normal star will little by little drift closer to the black hole and spiral inward and the X-ray emission will then be intense enough to detect over distances of many light-years.

In 1962, instrument-carrying rockets detected spots in the sky from which X rays originated. The most intense radiation so far discovered is called "Cyg X-1" because it is in the constellation of Cygnus (the Swan).

At the spot where Cyg X-1 is located, a visible star has been seen. This star, HDE226868, is 5.6 times as massive as the Sun and yet it is clearly circling another star larger still, one that is about 10 times the mass of the Sun. Such a huge star would ordinarily be bright, but it is not—it can't be seen at all.

Combine this invisibility with the fact of copious X-ray emission, and astronomers have come to suspect strongly that Cyg X-1 is an ordinary star circling a black hole.

The black hole of Cyg X-1 was detected because it happens to be large and fairly close to us (it is only a few thousand light-years away) and is part of a huge two-star system. Black holes that are smaller, or farther away, or not part of such a system, would go undetected. How many of them might there be altogether?

The Cornell astronomer Carl Sagan suspects there may be one black hole for every 100 stars. This would mean a billion black holes in each galaxy (including our own). The average distance between black holes might then be about 40 light-years, and any given star is likely to be within 20 light-years of some black hole or other.

But if black holes only absorb matter and never give it up, where does the process stop?

It doesn't. Eventually, everything in the Universe will tumble into one black hole or another (though the process may take trillions upon trillions of years, to be sure).

Some astronomers think that in the centers of galaxies, where the stars are very thickly packed together, any black hole that forms grows rapidly larger and that perhaps every galaxy has a huge black hole at its center, one that is slowly sucking into its insatiable maw everything on the outskirts.

We ourselves are some 30,000 light-years from the center of our own galaxy and between us and the center are vast dust clouds that hide the center. Perhaps that spares us the ability to detect the nearest of the giant black holes, and the one that is destined someday to be our swallower.

In 1963, astronomers discovered objects they call "quasars." These are very far distant from us, a billion light-years or more, and can be seen only because each is at least a hundred times as bright as an entire ordinary galaxy. Despite that brightness, they are quite small, much smaller than ordinary galaxies.

What can be so much smaller than a galaxy and yet, at the same time, be so much brighter than a galaxy? Can it be a black hole at the center of a galaxy? The black hole would be itself invisible, but the matter falling into it, on a huge enough scale, would produce an intensity of radiation greater than anything else we can detect. This is at least one possible explanation of the quasar.

What happens to matter when it plunges into a black hole? No one knows. There are no theories that astronomers can trust that suffice to describe conditions as extreme as those within black holes.

One speculation is that matter tumbling into a black hole undergoes fundamental changes in the characteristics of space-time about it. For instance, it may be that as the gravitational interaction grows more and more intense, the rate of time passage slows up.

Imagine yourself falling into a black hole under those conditions and yet capable of retaining consciousness. By the time you had fallen to within two miles of the center, you would be past the black hole horizon and unable ever to emerge again. By that time, though, your time sense will have slowed, and will be steadily slowing further, so that it would seem to you that it would take one year, perhaps, to fall an additional mile toward the center, leaving one more mile to go. By then, your still-slowing time sense will make it seem that it will take another year to fall an additional half mile toward the center, leaving another half mile to go; then another year to fall an additional quarter mile toward the center, leaving another quarter mile to go, and so on.

To someone viewing the fall from outside, the collapse would seem rapid, but to someone actually participating in the fall, it would take forever—it would be an endless fall down an infinitely long tunnel.

If the black hole were perfectly spherical and motionless, everything about it would be perfectly symmetrical and the infinitely far-off end of the tunnel would be in the same place in space that the original collapsing star had been.

If the black hole is not perfectly spherical, or if it is turning, or both, then, according to some suggestions, the endless tunnel would not remain in the same place in space. It would change position relative to the Universe and come out somewhere else—at some point enormously distant in space and time.

Matter emerging in this way at the other end of a particularly large black hole may do so with a vast explosion of energy. Light would pour out of it and we would see a "white hole."

It is possible that a quasar might not be a giant black hole but might be a white hole at the other end of a black hole? Might every quasar we see have an enormous black hole at the other end? And vice versa?

Sagan wonders if, perhaps, black holes may someday make trans-galactic travel possible. Traveling under ordinary conditions, we can't go faster than the speed of light, which means it would take four years or more to reach even the nearest star, 100,000 years or more to cross the galaxy, 2,300,000 years or more to reach the nearest other large galaxy, and so on.

We can't even conceive of anything that could keep a spaceship from being crushed to ultimate destruction if it enters a black hole, but perhaps advanced civilizations may learn of ways to resist the black hole forces, to enter one safely, to insulate oneself from changes in time sense, and to emerge many light-years away in an instant of time. By making proper use of black holes and traveling a minimum distance from one to the next, any point in the Universe might be reached from any other point in a reasonably short time.

Naturally, one would have to work out a very thorough map of the Universe, with black hole entrances and exits carefully plotted. The smaller black holes would be more numerous and therefore more important. One can only wonder how long a time exploring must continue before the entire Universe is mapped.

Clearly, it would be useful for a world to be near a black hole, since its citizens could then most easily travel to distant points. We might imagine some Cosmic Empire that is threaded through the network of black hole tunnels with its civilized centers located near the mouths. The planets nearest the tunnels might be a safe distance away, but nearer still would be enormous space stations built as home bases for the ships that thread their way through the black hole tunnel, coming and going.

These space stations can serve as power stations as well, since the energy radiated in the immediate neighborhood of a black hole can clearly be enormous. In fact, it might be the black hole energies, trapped by these space stations, some to be used on the spot, and some to be beamed back to home planets, that will form the real basis of the super-technologies of the Cosmic Empire.

We might even visualize space projects that consist of the moving of dead and useless matter into the black holes to increase the energy output (like fueling a furnace).

Would black hole energy be used as a source for Earth itself? Possibly, but by the time men had learned to make use of black holes, they would have become part of the Cosmic Empire and would have built themselves new homes on new planets more advantageously situated with respect to black holes. Earth would be left behind as a museum of humanity's origins, quietly rusticating in its hinterland.

Black holes can be of any size. Any mass, however small, that occupies zero space, has an infinite gravitation attraction in the immediate neighborhood. We can imagine a black hole that is very small indeed, perhaps with no more of a mass than an Earthly mountain, or even an Earthly molehill. Naturally such a "mini-black hole" will attract matter in its immediate neighborhood, pull it in, and slowly grow. It will not always remain mini.

But how can mini-black holes form? Black holes with masses of over 2.5 times the mass of the Sun form because of the pull of their own gravity. Smaller objects just don't have the pull.

In 1971, though, a Cambridge astronomer, Stephen

Hawking, suggested that in the process of a big bang, when vast quantities of matter are exploding all over the place, some different sections of the expanding matter might collide. Part of the colliding matter might then be squeezed together under enormous pressures from either side. Portions might be forced together and shrunken to the point where their own mounting gravitational interactions would keep them shrunk forever.

Perhaps similar mini-black holes are formed in quasars and in every white hole at the other end of a large black hole. If so, perhaps these mini-black holes litter space and form a very small, but perceptible, danger to us.

It might be that every once in a while such a mini-black hole (absolutely undetectable as long as it remained in deep space) may strike the Earth. Such an event may even have happened less than seventy years ago.

On June 30, 1908, in the Tunguska area in central Siberia, there took place what seemed to be a large meteorite strike. Every tree for twenty miles about the point of collision was knocked down. A herd of 500 reindeer was killed to the last deer (but, as it happened, there wasn't a single human death).

In later years, a search of the spot revealed no meteorite crater and no fragments of meteors. Scientists decided that the explosion must have taken place in the atmosphere.

The favorite theory for what happened is that a small comet struck the Earth. Meteorites are made of rock or of iron and hang together till they hit the Earth, even though they grow red-hot while passing through the atmosphere. Comets, however, are made of rocks and pebbles held together by icy materials that melt and boil easily when heated. The friction of the air may have boiled the comet so that the small rocks and pebbles, and the cometary structure generally, were torn apart with a loud and very forceful explosion.

But now the suggestion has been advanced that a mini-black hole was what struck the Earth on June 30, 1908. As it passed through the atmosphere, atoms by the millions fell in, releasing energy. It was the shock of that energy that knocked down the trees and killed the reindeer.

Eventually, the mini-black hole struck the Earth at an angle, and just passed through. Naturally, it sopped up more atoms and finally came out in the North Atlantic, where it would have to have produced a gigantic waterspout and explosion which, apparently, went unseen and unheard by man. It would then proceed on its way back into space, considerably larger than it was when it arrived, but still a mini-black hole.

There's no way of proving that that is what happened, except perhaps by waiting for another such event which will be better observed—though no one really wants that to happen.

It would seem, then, that black holes, both large and tiny, offer a variety of threats, from meteor-type collisions to ultimate and universal death—but also offer a dim and highly speculative hope of cosmic trade and empire.

Much more immediately important, they give us a chance of learning more about matter under very extreme conditions, and this may, in the end, give us key information as to how and why the Universe behaves as it does.

AFTERWORD

One thing leads to another. Having written the above article, but still smarting over my lack of awareness, I went on to write an entire book on black holes—Collapsing Universe, published by Walker & Company in 1977.

17 MAN AND COMPUTER

FOREWORD

Another of my enthusiasms is the computer. Back in 1939 (as I say in the essay below), I began to write stories about robots—which are a form of computer, housed in a man-like mechanical body. In my stories, robots were looked upon sympathetically. This was not completely new, but till then, the few cases of sympathetic robots had been drowned out in the numerous tales of evil Frankenstein monsters.

Without wasting time on modesty (which I know about because I have found the word in the dictionary) I can tell you that my own stories succeeded in changing the fashion in science fiction.

In the 1620s, the German astronomer Johannes Kepler was trying to work out tables that would predict the position of the various planets in the sky at any given time in the future.

It was not a new task. The ancient Babylonians had prepared such tables; so had the ancient Greeks; so had the medieval scholars. Some of them had been interested in the problem because it had astrological significance; some because it yielded an insight into the clockwork that ran the Universe.

Kepler had both interests at heart and he also had an enormously difficult task. To be sure, he had a record of careful observations of planetary positions in the past. He also had a new theory that he had himself devised—that the planets did not move around the Sun in circles, but in ellipses.

The calculations had to be done, then, on the basis of the new theory and they were not quite like those which had been done by earlier astronomers. Kepler was breaking new ground and he had to make a vast number of multiplications and divisions. It would take enormous time and at every step there would be the chance of arithmetical errors, so that he would have to check and recheck. He was already fifty years of age and life spans were not long in those days. Would he live to see the end?

Fortunately, a Scottish mathematician, John Napier, had devised a system of what he called logarithms. In 1614, he had produced long tables in which every number had another number, a logarithm, listed for it. If one added the logarithms of two numbers, the result was the logarithm of the product of the numbers. By adding (and using the table) one did the work of multiplying.

In the same way, if one subtracted the logarithm of one number from that of another, the result was the logarithm of the quotient. By subtracting, one did the work of dividing.

It is easier to add than to multiply, and easier to subtract than to divide, so the use of logarithms offered a tremendous shortcut in arithemetical calculations.

With great relief, Kepler snatched at the logarithm table and put it to its first important use in history. The task still took him a long time, but now he could finish the job in what was left of his lifetime. His planetary tables were published in 1627 and were the best ever produced up to that time. He died in 1630.

Logarithms are a computing device and nothing more, however. They added not one idea to Kepler's head. The notion of making the planetary orbits into ellipses instead of circles, of having the Sun in a particular position within the orbit, of having the planets move at different speeds according to their changing distances from the Sun—those were Kepler's ideas. Working out the consequences of the idea by arithmetical computation was merely detail, tedious and time consuming detail.

Even with logarithms, Kepler's task took years; and while Kepler, with his first-rate mind, was busy mumbling over his figures, he had no time to think up additional first-rate ideas.

You can be creative, or you can spend all your time adding and subtracting; you can't do both.

But imagine if Kepler could somehow have called on devices that were three centuries in his future. Suppose he could have made use of an electronic computer, one in which flashing electric currents, moving at the speed of light, mimic the rules of arithmetic, but follow those rules a hundred million times more quickly than the human mind can—and do so unerringly.

A couple of years ago, Kepler's raw data were indeed handed to an electronic computer, which was then programmed to do the kind of arithmetic it had taken Kepler several years to do, even with logarithms. The computer completed the task—in eight minutes!

Was the computer smarter than Kepler? Of course not. It was no more intelligent than a logarithm table is, but it was tremendously more complicated than a logarithm table is. The electronic computer is far more capable than a logarithm table would be to do just those mental tasks that are dull, repetitive, stultifying, and degrading, leaving to human beings themselves the far greater work of creative thought in every field from art and literature to science and ethics.

The coming of the electronic computer in the 1940s has introduced a revolution in the history of the human mind. It is, in some ways, a frightening revolution, one that takes away from human beings a burden to which we have so long been accustomed that it seems to have become almost a part of the human condition. Without it, some of us fear we would wither. —But that can't be so, for it is not dehumanizing to let a machine do dehumanizing labor. To let it do so will, rather, make human beings truly human at last.

We have the experience of an earlier, but less subtle, revolution of this kind behind us, and can judge from it.

Two hundred years ago, the energy of the inanimate world began to be tamed and made to do the physical

176

drudgery that till then had turned human beings into beasts. The steam engine, the internal combustion engine, the electric motor, did the work that till then straining muscles had done—did it faster, did it better—and made a measure of leisure the common possession of humanity rather than the exclusive prerogative of a tiny fraction of privileged people.

Did this first revolution dehumanize mankind? Was it better in the days when peasants, serfs, and slaves spent their brief, bitter lives in unremitting toil to scratch a semi-starvation out of the soil for themselves and their overlords? Can we long even for the pitiful splendor of those pre-industrial overlords, when that splendor would seem squalid by the standards of today's average?

Before science and technology had doubled the length of human life and tripled human health and comfort, what was there? Plagues. Minor infections that dealt death. Slavery. Endemic famine. Unremitting labor. —Who wishes a return to those days?

No society that has achieved the higher standard of living and the spreading affluence that comes with industrialization has ever voluntarily given it up.

Industrialization brought its problems, of course. The very advance in affluence and in medical prowess meant a rapid growth in population, an increasingly rapid consumption of resources, a quickly rising production of pollutants, a vanishing of space, and a decline of environmental quality.

This, however, is in part the result of the fact that the revolution relieved only the muscles of their degrading task of inhuman drudgery and left the mind alone. We could do more and more, but there was no way in which we could foresee more easily the results of what we were doing.

To be sure, the removal of unnecessary physical labor was, in itself, an advantage to the mind. It meant that each person had more time to think and that a larger fraction of the population could spend a lifetime in thought. More people could be scientists and engineers, so that the intensity and speed of the revolution grew greater, thus relieving still more people and forming a still stronger basis for still more thought and still more advance, and so on.

It meant that change came faster and faster, that improvement and novelty crowd upon improvement and novelty—

177

It becomes increasingly important to foresee consequences, and to foresee them more quickly, in order to weigh the relative advantages of taking this direction or that, of making or refusing to make this change or that, of taking or not taking this precaution or that. —And yet we can foresee, with our unaided mind, no better than we could in Stone Age times, and certainly no more quickly.

What is needed is a still further extension of the technological advance, one that can help us do our foreseeing, as electric motors help our muscles do their work.

Merely allowing more people more time to think has not proved to be enough. Combining numerous unaided minds and leaving all of them more time does *not* produce better answers more quickly.

The electronic computer, however, which can lift and *is* lifting the load of mental drudgery from the human mind, offers an answer to the problem. The product of the first revolution, it is initiating a second revolution greater and more rapid than the first.

The electronic computer cannot think *better* than the human being. Even the most complex electronic computer yet built is incredibly less complex than the three pounds of human brain with its hundred billion intricately interrelated cells—and with each cell, remember, possessed of an internal structure that is itself undoubtedly more complex than any computer.

The human brain, nevertheless, has its shortcomings. Though equipped to perform tasks of judgment, intuition, and creation, far beyond that of anything man-made, it is not particularly designed for speed. Speed beyond what it possesses was not vitally needed over the ages during which the human brain evolved.

The human brain, which evolved under the stresses of the needs and crises of a Stone Age society, has not noticeably improved in the instant of time (evolutionarily speaking) that has brought us from the Stone Age to our present high technology. It cannot, therefore, cope by itself with the infinitely greater problems of today.

No unaided human mind and no number of unaided human minds working in cooperation can solve the problems required to launch a spaceship and have it carry men safely to the Moon and back. Nor can they send an unmanned probe to a split-second rendezvous with distant Jupiter. The human mind could, indeed, solve the problems if it were given time enough, but the necessary time doesn't exist, and if it did the human brain would wear itself out with boredom. The electronic computer, with relays clicking at billionth-of-a-second intervals, *can* carry through the enormously repetitive operations required, and lacks the capacity to be bored doing so—or to make a mistake, either.

Nor can the unaided mind work out all the factors involved in using the weaponry of modern warfare—nor in dealing with the problems of bookkeeping involved in the collection of the nation's taxes—nor in straightening out the various questions concerning inventories, payrolls, sales, receipts, income, outgo, assets, debits, and moment-to-moment changes in all of these, in all our various industries and businesses.

The second revolution follows the same course as the first.

Once the machine took over the physical drudgery of mankind, there was no real desire to go back to the slave gangs and to pick-and-shovel construction. And once the computer took over, there is an equal unthinkability in going back to an army of clerks and bookkeepers.

Indeed, so much is done by the new devices that the pace could not be kept up even if every person on Earth were willing to pitch in as a replacement. The great modern works of man, the highways and bridges and buildings and dams could not have been built, without motors and machinery, in a hundred times the years they took even if everyone on Earth were conscripted into unremitting back-breaking labor. And the great computing decisions of man cannot now be made in time by the calculating minds and racing pens of clerks and secretaries and bookkeepers even if every human being were forced onto a stool with an open ledger before him.

The machines of the first revolution gradually became smaller and more versatile. Whereas at first they were used in factories only, the time arrived when private individuals could each have his motor-driven devices, from an automobile down to an electric toothbrush.

In the same way (and much more rapidly), where computers were originally shambling monsters that only universities and government installations could possess, they have become steadily smaller, more versatile and flexible, until they have grown useful and even indispensable to the smallest businesses.

Indeed, it is now fashionable for individuals to carry

relatively cheap, shirt-pocket-size computers to do the routine work of adding up bank balances, shopping lists, and tax deductions.

Among the great tasks that the computer will be asked to undertake is that of determining the proper human strategy for dealing with the problems raised by the first revolution.

Now perhaps we can plan ways of accomplishing great change while also foreseeing with useful clarity the consequences of those changes, and therefore so guiding them as to achieve what we want and expect, rather than stumbling into what we don't want and didn't expect.

It will not be easy. We must learn to ask the proper questions; to devise programming techniques that will enable computers to produce useful answers to the questions in a reasonably short time; and to devise computers complex enough for the purpose.

But won't we be able to do all this? Computers, at least, are steadily advancing in complexity, versatility, and usefulness. We already have chess-playing computers, music-writing computers, language-translating computers. Undoubtedly we will have computer-designing computers and perhaps even question-designing computers. It is conceivable that we may reach the point where we can ask computers what questions we ought to be asking them, and where we can program them to design better programs.

But if we can accomplish that, and if we can use the second revolution to cancel out or, at least, minimize the disasters brought about by the muscle-not-mind advance of the first, what certainty do we have that, as the years pass, the second revolution will not itself bring about new and unprecedented horrors? After all, every solution is likely to spawn new problems of its own.

We may indeed develop a new society in which all seems well and in which a veritable Utopia will have been achieved, but it will have been at the expense of relying on our computers, and might not this very reliance become itself a danger? Will computers become so complex and versatile that they will develop an intelligence approaching or *surpassing* that of the human being? Might not computers then "take over"? Might they not kill us off and replace mankind as lords of the Earth?

Actually, I have myself considered this problem in a series of science fiction stories I have written over the past thirty-five years that concerned computerized man-like devices called "robots."

My own solution was to suppose that no matter how advanced computers might become, they would always be designed by men and that men would, in their own interests, build certain ineradicable safeguards into their creations. I expressed these as what I called "The Three Laws of Robotics," which go as follows:

1. A robot may not injure a human being nor, through inaction, allow a human being to come to harm.

2. A robot must obey the orders given it by human beings except where such orders would conflict with the First Law.

3. A robot must protect its own existence as long as such protection does not conflict with the First or Second Laws.

I won't say that precisely these rules will govern the computers of the future (though, if that should come to pass, and I were to live to see it, I would not be surprised), but it seems unquestionable to me that some sort of precaution will be included.

The notion of built-in precautions is, after all, as old as technology, and it is not likely that we'll abandon the practice in the future. Knives have handles, swords have hilts, triggers have safety catches, elevators have automatic stopping devices, buildings are made of fireproof materials—and computers will have something equivalent to the Three Laws.

But computers are the most complex devices ever constructed by man and are becoming steadily more complex. Can we trust them to remain bound and inhibited? Might they not get away from us despite our best efforts to prevent it?

And even if they don't and if they remain our faithful servants, might not our ever-increasing dependence on them wither our own mental initiative and everything else that is of human value in our minds?

Yet need we fear this? Can we not suppose that as computers advanced, the human mind might advance, too?

Until now, in the history of life, evolution has proceeded through random gene changes and through a brutal and not very discriminating natural selection. Evolution has therefore been an exceedingly slow process and has been riddled with blind alleys.

We have now reached the stage, though, where the advance of genetic engineering makes it quite conceivable that we will begin to design our own evolutionary progress —and will be stopped from doing so not by any strictly technical difficulty but from the overall ethical problems of deciding how we want to change, what we intend to become, where we intend to go, who should be subjected to the effects, why the change should be introduced, and in what way we should decide on the hows, whats, wheres, whos, and whys.

These are questions of a new range of subtlety which we may be able to formulate in such a way that the advanced computers of the future could tackle them profitably. With the advice of the computers and, moreover, with the example they will offer of a functioning intelligence whose detailed structure we have designed and therefore (it is to be hoped) understand, the evolution of a steadily more intelligent human race will proceed in tandem with the evolution of a steadily more intelligent computer.

Man *and* computer may go forward together, mind-in-mind, someday. Man's mind, with its accent on intuition and creativity, and the computer's mind, with its accent on speed and precision, can together become a combined problem-solving tool far greater than either alone.

And if that should come to pass, I can see no bounds and no final horizons on the path to the infinite.

AFTERWORD

I don't delude myself into thinking that in essays like the above I can single-handedly allay human fears of computers. I try to, nevertheless. After all, considering what human beings do and have done to human beings (and to other living things as well) without computers, I can never imagine what the devil people think computers can add to the horrors.

C

THE
FUTURE

18 THE BIG WEATHER CHANGE

FOREWORD

I must admit that while I am always ready to write articles, I am a bit readier to write them for TV Guide than for many another magazine. The articles for TV Guide cover such a wide range of possibilities (after all, consider how miscellaneous a medium television is) that I am very often breaking fresh ground, for myself at least, in my work for it. I hadn't ever written about the future of weather, for instance, until TV Guide asked me to.

During the last million years, the world has seen huge oscillations of warm and cold. Sometimes the Arctic Ocean is open water and then it supplies water vapor which falls on the surrounding land areas as snow. If there is then a small drop in general temperature for a prolonged stretch of time, not all the snow that falls each winter will melt each summer.

The snow then accumulates from year to year and squeezes down under its own weight to form glaciers. The glacier ice reflects Sunlight more efficiently than bare ground does and cools the Earth further, so that still less of the snow that falls in the winter is melted, and the glaciers advance southward.

Eventually, the temperature drops to the point where the Arctic Ocean freezes over and the supply of water vapor is cut off. Less snow falls, so that the summer melting becomes more effective and the glaciers begin to retreat. The retreat reduces the ice cover, allows the Earth to warm, and accelerates the retreat further. When it grows warm enough for the ice on the Arctic Ocean to melt, it starts all over.

In the last million years, the glaciers have advanced four times and retreated four times. There have been long winters, ten of thousands of years long, with glaciers reaching to the Ohio River and to the Ukraine, so that 30 percent of Earth's land surface was covered by ice. In between there have been even longer summers, with the weather somewhat warmer than it is today.

The last maximum ice cover was at about 16,000 B.C. and the final retreat began about 8000 B.C. It wasn't till 3000 B.C. that the ice was confined to Greenland and Antarctica as it is today.

The coming and going of the glaciers have not, in the past, represented great disasters. Northern lands may become ice-covered, but the storm tracks move farther south, so that the great semitropical deserts disappear and the Sahara, for instance, becomes a lush grassland. Then, too, so much water is tied up in the huge glaciers that the sea level drops by as much as 300 feet and large areas of the continental shelves surrounding the continents are uncovered and elephants can graze on them for thousands of years.

When the glaciers retreat, the continental shelves are flooded and the deserts reappear, but the northern lands are open and fertile once more.

The glaciers advance and retreat very slowly—only a couple of hundred yards a year on the average—so there is plenty of time to make the adjustment. The ancestors of man lived through all the Ice Ages.

And what about the future? Will the glaciers return?

They may if the Earth's temperature changes slightly. If it rises a little, that may melt the Arctic Ocean's ice cover and trigger the glaciers. If it falls enough, that, too, may serve as a trigger even if the Arctic Ocean stays ice-covered.

As it happens, man's twentieth-century industrial activity is making temperature changes. All his burning of coal and oil is increasing the carbon dioxide content of the air very slightly and that extra carbon dioxide acts as a heat trap to raise Earth's temperature slightly. He also produces dust and other particles and these tend to reflect Sunlight back into outer space, thus cooling the Earth.

From 1900 to 1940, the carbon dioxide effect predominated and Earth's temperature rose slightly. Since 1940, the dust effect took the lead and Earth has been cooling. In each case, the rate of temperature change, whether warming or cooling, has taken place more rapidly than ever before in Earth's history as far as we know.

The warming wasn't enough to trigger a new Ice Age, and neither has the cooling been—so far. But Earth's temperature is still going down now, and the trigger may yet be pulled.

On the other hand, if we clean up the air so that the carbon dioxide effect is no longer balanced by the dust, *all* Earth's supply of ice may melt, even the great ice cap of Antarctica, which has lasted through all the previous summers between the Ice Ages. In that case, the sea level would rise some 200 feet above its present level, and the coastal plains, in which much of the world's population lives, would be drowned by in-creeping surges of water that would finally reach the twentieth story of the Empire State Building.

A fifth big weather change, in either direction, would be far more of a catastrophe than the first four, for the following reasons—

1. The change would be much more rapid than ever before. Thanks to man-made temperature swings, what took tens of thousands of years in previous cases might take only centuries this time. Adjustments would have to be that much more rapid.

2. Man's population has risen. When the last period of glacial advance began, there were perhaps four million human beings on Earth. Now there are four thousand million. The adjustment would be enormously more difficult.

3. Human beings and their ancestors had a hunting economy in previous periods of glacial advance and retreat. They could, without trouble, follow the glaciers north and south. Nowadays, mankind has a farming

and industrial economy; and farms, mines, factories, and cities are hard (and sometimes impossible) to move.

Can we avoid a big weather change, perhaps? Yes, if one of two things happens. If (a) for some reason the world temperature happens to remain stable and does not either rise or fall sufficiently to set off a drastic change, or (b) if scientists learn enough about the weather to devise methods of controlling the trigger—then we can escape.

Maybe (a) it will, or (b) we can. Otherwise, we're in trouble.

19 TIGHTEN YOUR BELT

FOREWORD

In 1973, a new magazine called Good Food *was begun by the wonderful people of* TV Guide. *Merrill Panitt, the debonair editorial director of the latter, was to perform the same function for the former. We had lunch together and conspired to have me write an article for the new magazine, since my devotion to good food is legendary. The following is the result.*

The future depends, as it always has, on the present; and the present consists of what the "doom criers" (myself included) have constantly predicted, and what most people have steadfastly refused to accept—an end to indiscriminate growth; that is, growth for growth's sake.

The fact is that the energy crisis, which has suddenly been officially announced, has been with us for a long time now, and will be with us for an even longer time. Whether Arab oil flows freely or not, it is clear to everyone that world industry cannot be allowed to depend on so fragile a base. The supply of oil can be shut off at whim at any time, and in any case the oil wells will all run dry in thirty years or so, at the present rate of use.

New sources of energy must be found, and this will take time, and is not likely to result in any situation that will ever restore that sense of cheap and copious energy we have had in the times just past. We will never again dare indulge in indiscriminate growth. From here on in, for an indefinite period, mankind is going to advance cautiously, and consider itself lucky that it can advance at all.

To make the situation worse, there is as yet no sign that any slowing of the world's population growth is in sight. Although the birthrate has dropped in some nations, including the United States, the population of the world, generally, seems sure to pass the six-billion mark as the twenty-first century opens and perhaps even the seven-billion mark. The food supply will not increase nearly enough to match this, which means that we are heading into a crisis in the matter of marketing and distributing food.

Taking all this into account, what might we reasonably estimate supermarkets to be like in the year 2001?

To begin with, the world food supply is going to be steadily tighter over the next thirty years—even here in the United States. By 2001, the population of the United States will be at least 250,000,000 and possibly 270,000,000, and the nation will be hard put to expand food production to fill the additional mouths. This will be particularly true since the energy pinch will make it difficult to continue agriculture in the American high-energy fashion that makes it possible to combine few farmers with high yields.

It seems almost certain that by 2001 the United States will no longer be a great food-exporting nation and that, if necessity forces exports, it will be at the price of belt-tightening at home.

This means, for one thing, that we can look forward to an end to the "natural food" trend. It is not a wave of the future. All the "unnatural" things we do to food are required to produce more of the food in the first place, and to make it last longer afterward. It is for that reason that we need and use chemical fertilizers and pesticides, while the food is growing, and add preservatives afterward.

In fact, as food items will tend to decline in quality and decrease in variety, there is very likely to be increasing

190

use of flavoring additives. Until such time as mankind has the sense to lower its population to the point where the planet can provide a comfortable support for all, people will have to accept more artificiality and not less.

Then, too, there will be a steady trend toward vegetarianism. A given quantity of ground can provide plant food for man; or it can provide plant food for animals which are later slaughtered for meat.

In converting the tissues of food into the tissues of the feeder, there is up to 90 percent used for reasons other than tissue maintenance and growth. This means that 100 pounds of plant food will support 10 pounds of human tissue—while 100 pounds of plant food will support 10 pounds of animal tissue, which will then support 1 pound of human tissue. In other words, land devoted to plant food will support ten times as many human beings as land devoted to animal food.

It is this (far more than food preferences or religious dictates) that forces overcrowded populations into vegetarianism. And it will be the direction in which the United States of 2001 will be moving—not by presidential decree, but through the force of a steady rise in meat prices as compared with other kinds of food.

This, in turn, will come about because our herds will decrease as the food demand causes more and more pastureland to be turned to farmland, and as land producing corn and other food for animals is diverted to providing food directly for man. And in the suburbs, the lawns and flower gardens will be converted into vegetable plots as during World War II. They will be "survival gardens," rather than "victory gardens."

Another point is that it is not only energy that is in short supply, thanks to a world population that has increased madly throughout the twentieth century and a philosophy of endless waste that has governed American thinking in particular. A shortage of oil means a shortage of plastics; a shortage of electricity means a shortage of aluminum. We are also experiencing a shortage of paper and of most other raw materials.

This means that, for one thing, there will be a premium on wrapping. Our generosity in wrapping, bagging, and packaging will have to recede. There will have to be an at least partial return in supermarkets to the old days when goods were supplied in bulk and ladled out in bags to order. It may even become necessary to return bags, as we once returned bottles in days of yore, or pay for new ones.

A decline in per-capita energy use will make it necessary to resort to human muscle again, so that the deliveryman will make a comeback (his price added to that of the food, of course). This will make all the more sense since energy shortages will cause unemployment in many sectors of the economy and there will be idle hands to do the manual work that will be necessary.

It would make far more sense, after all, to order by phone and have a single truck deliver food to many homes, than for a member of each home to drive an automobile, round-trip, to pick up a one-family food supply.

To be sure, it will not all be retrogression. Even assuming that Earth is in a desperate battle of survival through a crisis of still-rising population and dwindling energy reserves, there should still continue to be technological advance in those directions that don't depend on wasteful bulk use of energy. Ther will be continuing advances in the direction of "sophistication," in other words.

Most noticeably, this will mean a continuing computerization and, where possible, automation of the economy.

By 2001, we can imagine devices that will make the phoned-in order more versatile and more precise. We might imagine a centralized supermarket catalogue, issued once a year, that lists, with description, all that is in stock. Brand names are likely to be less important as time goes on, and they will not be emphasized. If you want peas, you will get peas in, very likely, a plain package. minimally marked—or just a bag, filled by a computerized device responding to the size of your order.

Furthermore, you cannot expect unlimited quantities of an unlimited variety of goods. We can imagine weekly pamphlets, listing out-of-stocks and in-stocks, together with updatings in prices and descriptions—and, very likely, setting a maximum size for orders in some cases.

There would be codes for each item and for the quantity desired and this could be punched in on a code board attached to the telephone, together with a credit-card number. The completed punch-in could be typed out on a strip of paper as it is formed, so that the result could be carefully cross-checked with the catalogue, since you wouldn't want to get spinach when you ordered chocolate cake. Errors can be adjusted; there can be deletions and additions; and when you are satisfied, a final punch would transmit the order to the supermarket. You will retain a record, against which you can compare the order when the deliveryman finally brings it.

It all sounds tedious, perhaps, but it is not as time-consuming, or as energy-consuming, as personal shopping. And though you would be deprived of the chance to squeeze your own tomatoes and thump your own canta-loupe, it is the question of energy consumption that counts.

A thoroughly computerized and automated supermarket could be open on a twenty-four-hour-a-day, all-days-a year basis. There would be ample energy for that if the energy consumption of personal shopping in myriads of small stores (all of which will be disappearing rapidly) is removed. This would make it possible to supply emergency numbers for special delivery of limited amounts of certain staples at any time.

It is also very likely that the whole thing would work best in a cashless economy. The charge would be toted up and subtracted from your cash reserve and you would know at the time of ordering—or, for that matter, at any time you choose to check—the status of that reserve. Go-ing into the red would be troublesome, for, if you did, it would become increasingly difficult to be fed—as is true right now, for that matter.

In one respect the supermarket trend would be toward the supplying of staples in bulk; but in another there would be a tendency toward prepared food—saving the energy required for individual preparation in millions of kitchens in which the supply of cooking gas or electricity will be increasingly precarious.

This may take place in a variety of ways—frozen foods, as we have now, which will, however, become less popular since they require considerable energy for heating; or ordi-nary cooked food, prepared to order, and delivered with reasonable speed. Sandwiches of one sort or another may be the most popular item in the supermarket; and may of-fer the greatest variety and the most imaginative creativity in the field of food. (They may also offer the major supply of what meat there is.)

The restaurant industry in general will very likely suffer a decline in the next generation and they will be most eco-nomical and successful in close association with the super-market. There will be savings in eating on the spot, assuming you can get to the supermarket without undue expense, and the "super-teria" may be the restaurant of the future.

Of course, the future has a future of its own, and it is possible that by 2001 there will be a move toward using the world of microorganisms as a new cheap source of protein and other important nutrients.

Microorganisms—algae, fungi, even bacteria—can live on materials not suited for food otherwise and, under favorable conditions, can grow and produce edible nutrients far more rapidly than larger forms of life can.

"Micronutrients" could be refined and eventually routinely added to flour, let us say, so that fortified bread, sold through the supermarkets, might serve as the major supply of protein and vitamins for the underfed of the world.

All this may not sound as though the year 2001 is going to be a great one for gourmet dining—but that's because it just won't be. Good food and careful preparation to individual taste may still exist, but fewer people will be able to afford it, and they will be able to do so a smaller percentage of the time.

This is regrettable, but the blame rests with a human foresight that was too slow to see the inevitable consequence of overpopulation combined with the waste of resources. And yet, if mankind learns its lesson, and acts as promptly as it can to limit population and establish a conservationist economy, progress may resume and a world of reasonable plenty for a reasonable population may return—perhaps by 2101.

AFTERWORD

Good Food *was a delightful magazine, made to order for sales in the supermarkets. By a stroke of rotten luck, however, almost immediately after the first issue appeared, there struck the Arab oil boycott (which I mention in the article). Up went the price of paper, ink, and so on, while the magazine's fees for advertising were, for the period, fixed by their initial guarantees. It couldn't make it, therefore, and ceased publication.*

20 AMERICA— A.D. 2176

FOREWORD

The year 1976 was the bicentennial year of the United States, as everyone knows, and it was a bonanza year for me in terms of bicentennial articles.

What made it a little difficult, though, was that a surprising number of newspapers and magazines came up with roughly the same idea—looking into the American future rather than back into the past—and all of them seemed to turn to me.

As upon this day—July 4, 2176—we look back over the four-hundred-year history of the United States of America, we are astonished at the differences each century has marked.

In the first century of the nation's existence (1776-1876), the United States was an isolated society building its institutions as independently as possible of the swirl of international complications that then centered in Europe.

In the second century (1876-1976), it became first a

195

great power, then the greatest power, and, for a period of time, it dominated the planet both militarily and economically.

In the third century (1976–2076), it was the embattled remnant of an apparently collapsing world, trying to restore order and to build a society on newer and saner foundations.

Finally, in the fourth century (2076–2176), it is part of that sane world, one that still bears the marks of the Catastrophe of a century and a half ago. We no longer have true wilderness, for instance, as a result of the desperate struggle for food in the last decades of population increase. The number of species of life still in existence, particularly among the larger and more splendid forms, has decreased substantially. Parts of both land and sea have not, even yet, regained the full flourish they possessed two centuries and more ago.

But there is this—the Turnabout, which was reached in 2050, has persisted, with occasional falterings, for over a century, and it seems reasonable to believe now that, barring unforeseen catastrophe, our society may continue to live and develop into the indefinite future.

Though the United States of America is no longer an independent nation in the old sense, and is part of the Global Federation (GF), the American people have rallied to celebrate their tetracentennial with undiminished enthusiasm.

Whatever political and economic changes the world has seen in recent centuries, the American tradition of popular participation in the political process, and of the protection of the individual against the group, remain. The United States still offers its easygoing attitudes in these respects as examples to those regions of the GF that retain more autocratic control over their local populations.

The United States also remains the most innovative and creative of the regions. It leads in scientific advance and in the exploration of space; in the extent of computerization and of automation; and in the flexibility of its educative procedures. If there is no longer an unhealthy disparity of living standards between the United States and much of the rest of the world, there are still a number of intangible benefits that accrue to Americans because they are living where they do.

The reason for this rests on two chief historical facts. First, the United States happened to be the last nation on

Earth to maintain a food surplus into the period of Catastrophe (1990–2050). It was also able to achieve early, and with minimal desperation, a low-birthrate society and a controlled population decrease. It was therefore less battered than most of the world was by the Catastrophe.

Our present population of 165 million is a hundred million less than our maximum population, achieved a century ago, but this is not a drastic drop and it was achieved without undue increase in the deathrate. Consider the Bangladesh region at the other extreme, however, where a population decrease to some 18 percent of the maximum was brought about almost entirely by a rise in the deathrate.

Our decreased population, brought about almost entirely by a lowered birthrate, has perforce made us undergo an important change in age pattern. The proportion of those over forty-five has been increasing steadily and, according to the census of 2170, half the population of the United States is now over forty-five. Except for a few of the European regions, this pattern change has nowhere been as extreme as here.

Two centuries ago, such a change in age pattern would have seemed catastrophic in itself. So large a proportion of the population would have been retired from the labor market, would be drawing pensions, would be overloading medical facilities, that the withered remnants of productive youth would have proved unable to stand up under the strain.

That this has not happened shows how society has changed since the Turnabout.

Medical advance has centered increasingly on gerontology, so that our aging population is healthier and more vigorous than they would have been in past centuries. It is now quite common for men and women to maintain their vigor to the century mark.

Then, too, the advance of computerization and automation has lifted the burden of dull and repetitive tasks that made up so much of the apparently necessary work of the world. The social organism can now be made to run with much less effort and in much less time. There is no concern these days about the effects of lowering the percentage of "productive youth."

But it is the revolution in communication that has, perhaps, been the most remarkable change of the period since the Turnabout.

The development of communications satellites interconnected among themselves and with the Earth by modulated laser beams has supplied every person in the United States (and very nearly every person in the world) with his own wavelength.

The new communication gives each of us the ability to speak (and see) anyone else on Earth at will and with minimal delay. Each one of us can now see or hear, at will and with minimal delay, any document, periodical, or book in the common domain that is maintained by the central computers at the Library of Congress, the British Museum, the Leningrad Library, or any of the other great collections, and to obtain a facsimile of it, if desired.

Each one of us is able to hold conference at any distance and in all practical numbers, by way of holographic televised images, indistinguishable from physical presence as far as sight and sound are concerned. Each one of us is able, if we are professionally involved, to control affairs at offices, factories, or farms through telemetered and television-monitored automation.

The result of all this is that travel, except for pleasure, has declined very sharply. It is no longer necessary for human beings to congregate into dense masses in order to be near their centralized places of work, or to enjoy the cultural advantages once available only at places of high-density population. The population decline of the last century has also seen a decentralization, therefore, and a steady fading of the city.

The new communication has produced a revolution in education as well. Elective instruction by television cassette and other devices has for decades outweighed, in importance and use, mass instruction in the classroom. People at any age can now receive instruction in private, at home, at will, at one's own rate, in subjects of one's own changing interests.

As we all know, of course, the privilege of learning involves the duty of teaching, and more Americans involve themselves in one phase or another of the teaching profession—the programming of computers, the preparation of cassettes, the development of new skills and knowledge—than in any other.

198

None of this would, of course, have been possible without a stable and long-enduring energy supply.

The United States, during its first century, was powered by burning wood and coal; and during its second century, by burning coal, oil, and natural gas.

During the third century, there came the fusing of hydrogen, something which made the Turnabout possible. It was through hydrogen fusion that a means for efficient cycling of resources was first developed. Wastes, which had previously been helplessly dumped on land and sea to make ugly and to poison the world, now passed through a fusion-powered "plasma torch" which reduced everything to its elements again and allowed a fresh start. It is for this reason that there are virtually no mines in the United States today. Our wastes are our mines.

And now, during the fourth century, our chief energy source is the radiation of the Sun as collected by the Solar Power Stations in space and transmitted to Earth as microwaves. This new and copious source of energy has coincided with advances in our knowledge of photosynthesis to make it possible for us to grow food with remarkable efficiency on relatively small farms, and to convert much of the Earth into parkland and nature preserves.

The mention of the Solar Power Stations is a reminder that the United States today, as well as the whole GF, depends on space for its prosperity.

And for a new stage of human expansion, too. The Global Federation, which, when it was founded, was intended to include all human beings within itself, does not, as of this day, do so any longer, for it no longer includes the space colonies in the Lunar Orbit.

Those space colonies are of peculiar interest to Americans since the first one built, soon after the Turnabout, was constructed chiefly through American effort, and was populated chiefly by American colonists who went on to construct the first Solar Power Station. American engineers also contributed notably to the technique of comet-trapping, whereby the colonies needed no longer to rely on Earth for supplies for the light elements—carbon, hydrogen, and nitrogen—so necessary for life and yet absent from the general space-colony quarry, the Moon.

At the present moment, five small comets, each rich in carbon, hydrogen, and nitrogen, are preserved within the lips of the Moon's polar craters, where they are never exposed to the Sun and therefore do not lose their precious volatiles.

Partly because of this peculiar interest in the colonies, the United States has been a leader, here within the GF, in the cause of colonial independence. It has argued forcefully and steadily that whereas it makes sense for Earth's interconnected and unitary globe to be under a single overall government, that government cannot properly rule over the multiplicity of smaller worlds, which have problems and interests peculiarly their own.

It is a source of deep satisfaction to almost all Americans, then, that the tetracentennial of our own freedom will witness a new birth of freedom elsewhere. A Lunar Federation (LF), independent politically, though not economically, of the Global Federation, is being officially established on this very day, July 4, 2176.

The population of the Lunar Federation, 750,000, is but a small fraction of Earth's human load of two billion. However, the population of the LF may be expected to grow rapidly, while that of the GF is stable.

And what of the future? What of the pentacentennial?

Physicians are increasingly optimistic that the aging process can be arrested indefinitely once enough is learned about the physiological and biochemical details of the process. Perhaps, then, at the pentacentennial, Earth will consist of a population that is potentially immortal—or at least one where each person can live until such time as accident or personal desire dictates an end to life.

The birthrate must then fall even lower, but it seems quite likely that by the pentacentennial, if not sooner, the technique of ectogenesis will have been perfected and women will no longer serve as personal incubators for fetuses. Instead, fertilized ova will be contributed to special clinics where fetuses can be grown in artificial uteri, with their genetic combinations mapped in the process and with deficiencies eliminated.

There is no question but that our new LF will greatly expand in the course of the next century. The Lunar Orbit is already crowded and humanity's "manifest destiny" points outward to the asteroidal region, where there is ample room and much raw material to be dismantled and used. In fact, where the Moon is very poor in the light elements most essential to life, the large asteroids are, in some cases, rich in them.

Indeed, it is not at all unlikely that eventually some colonies will choose to outfit their small worlds with an advanced space drive (beyond the simple and rather inefficient nuclear drives in use today) and leave the Solar system altogether in order to wander indefinitely among the stars.

Such a deep penetration of the Universe would mark the beginning of a new era of humanity, one that would far dwarf all that went before and would mark, perhaps, our entry onto a new and immense stage already occupied by other intelligences as well.

And it may be that the first starship will set forth on its travels on the Pentacentennial Day, July 4, 2276.

AFTERWORD

My own celebration of Bicentennial Day, incidentally, was a delightful one, entirely owing to the foresight of my wife, Janet. As soon as she heard, months before, that there was to be a parade of tall sailing ships up the Hudson, her Viking blood stirred (her ancestry is three-quarters Scandinavian and one-quarter English). She reserved a hotel room immediately overlooking the Hudson, my brother and his wife joined us, and we had a swell time.

21 THE COMING DECADES IN SPACE

FOREWORD

The next essay is another one of those in which I preach the desirability and even necessity of space exploration. This one, appearing in Boys' Life, *was written for youngsters, yet despite that (or perhaps because of that) I seem, as I read the article over, to be making my points more clearly than in most of my exhortations.*

We have reaeched the Moon!

In 1957, mankind launched its first satellite into space. That first satellite, unmanned and small, circled Earth, barely skimming the main body of the atmosphere. In 1969, a large, three-man vessel traveled safely to the Moon, 237,000 miles from Earth, deposited the first human beings onto the soil of an alien world, then brought them safely back.

So much for what can be done in a dozen years. What will we do in the next two and a half dozen?

Reaching the Moon by three-man vessels in one long

bound from Earth is like casting a thin thread across space. The main effort, in the coming decades, will be to strengthen this thread; to make it a cord, a cable, and, finally, a broad highway.

Can we stand the expense of doing this?

What expense? The manned space program has cost the United States $24 billion, but it was $24 billion spent right here in the United States (not on the Moon). It went partly into the development of many materials, devices, and techniques that will have, and already do have, applications here on Earth. It went partly into laying the groundwork for future advances that will be less expensive and will make greater returns.

Right now, the average American is paying $15 per year on the space program. He also pays $35 per year on alcoholic beverages, $17 per year on tobacco, and $16 per year on cosmetics. Ask yourself which of these is the biggest bargain. Is smoking oneself into lung cancer, for instance, more exciting and rewarding than landing on the Moon?

To make the landing more than a sort of glorious feat and to bring it into the realm of the commonplace, we are going to need some base in between Earth and Moon.

We will need a large manned satellite, orbiting the Earth permanently at a distance from us far less than the Moon. The Soviet Union has been gearing its space program toward the establishment of such a "space station," and the United States plans to have its first space station in orbit in 1973. The American space station will be powered, in part, by Solar batteries and is planned to include a large reflecting telescope.

The telescope indicates the most immediate use of the space station. It will be an observation post. Though the station may be just a few hundred miles up from the Earth's surface, it will nevertheless be millions of miles closer to the rest of the Universe than we are down here.

Why? Because it will be outside the atmosphere.

Our atmosphere is a crowded mass of molecules that lets hardly anything through. Most of the visible light gets through and most of a band of shortwave radio radiation called microwaves. Everything else is absorbed. What's more, the water vapor in the atmosphere forms mist, fog, and clouds that cut off the visible light, too, frequently.

There are some areas in deserts and on plateaus where

203

the air is clearer, but even there temperature differences from one part of the atmosphere to another bend light beams, make stars shift position, blur planet surfaces. And even in the most isolated spots, the dust in the air is increasing, distant light from man-made cities grows brighter. It becomes harder to see.

Every year the atmosphere becomes less transparent, harder to look through, and it is a space station that will take us beyond all that. The vacuum of space is absolutely clear and a telescope with no air between itself and the stars is an astronomer's paradise.

One thing we would most clearly see from a space station (and something we cannot see at all from Earth's surface) is a large swath of the Earth taken as a whole.

The Earth can be mapped from space with an instant accuracy available by no other means and the maps could always be kept up to date. It would be an extraordinary map, that would show details no other map could. It would show the character of the ground and from small changes we could deduce the location of mineral and oil deposits. We could study geologic faults more precisely and learn to predict landslides and earthquakes. We could follow the changing patterns of vegetation, study the progress of plant diseases, make farming more efficient.

It may be only by studies from space stations that we will learn enough about the Earth as a whole to make real progress in studying ecology; in studying the worldwide relationship of life forms to each other and to the environment.

Nor is it only the Earth that can tell us about the Earth. From a space station, we can study the Sun as it has never been studied. We can study *all* its radiation in detail and in full, and not just what radiation gets through our thick and dusty atmosphere.

We will better understand the amount and kind of energy that the Earth receives, and the amount and kind it radiates away from different portions of its surface. We will understand its weather systems better and learn, perhaps, how to introduce controls that will make the environment better for man and for life generally.

We will better understand the cosmic-ray particles and the neutrinos that fill all of space, and learn what they may tell us of the inner workings of stars and the evolution of galaxies. We may discover new particles that may reveal new mysteries undreamed of. —But will these exotic new facts about these particles be useful?

Certainly! Knowledge is *always* useful; if not at once, then someday.

Knowledge is all one piece. No matter where we increase the knowledge; no matter what we learn about the Universe; no matter how far-off and unimportant it all seems, it could have ways of helping us right here on Earth.

For instance, what could seem more unimportant to the average human being than the fact that quadrillions of miles away from us there are dust clouds in space?

Well, by studying the kind of light emitted from the dust clouds and the kind of light they absorb, astronomers have been able to deduce what kind of atoms exist there. Mostly, they are hydrogen and helium atoms, but with some oxygen, carbon, and nitrogen atoms thrown in. These atoms are spread out so thinly that it didn't seem likely that very many would just happen to collide and stick together.

But then in the 1960s, microwaves were studied instead of just light waves. That made more delicate decisions possible about any atom combinations that could exist there. In 1963, oxygen-hydrogen combinations were detected.

Then, beginning in late 1968, more and more atom combinations were found; two, three, four, even six atoms in combination have been located in those clouds. To detect them requires the most delicate work with radio telescopes.

In space, though, away from interference by Earth's man-made radio waves, and with the ability to observe waves of all sizes and lengths, it would be possible to study the substances in the dust clouds much more carefully.

And if we can do that—well, remember that the Sun and the Earth were formed out of such dust clouds. We might end up with a better idea of the chemistry of the Earth as it was first formed, and of what chemical changes must have taken place thereafter to form still more complicated molecules.

If we have a better idea of the starting chemistry of Earth, we may be able to conduct more meaningful experiments that will allow us to deduce the pathway to life. We may be able to construct the very simple systems

205

that would begin to hint of life. From them we may be able to learn some of the basic behavior of cells—behavior that is now masked by all the complicated additional molecules and systems that have been added by three billion years of evolution.

In other words, a clear line of research may extend from the constitution of the dust clouds (as determined from a space station) to a better understanding of the fundamentals of life.

Naturally, a space station will have to be kept supplied. It may need repairs. Scientists and engineers who have served their hitch will have to be taken off and replaced with others. There will have to be a continual line of ships shuttling between Earth's surface and the space station.

A space shuttle is now being planned; a double vessel, with the smaller (the orbiter) being carried pickaback by the larger (the booster).

Both the booster and the orbiter will be equipped with rocket engines carrying fuel and oxygen for use in deep space, but each will also be equipped with jet engines that make use of atmospheric oxygen. Each will be equipped with wing and tail elements to help them navigate in the atmosphere.

This represents an important revolution in space flight.

Until now, nothing sent into space has ever been used again. Objects in orbit stay in orbit forever, or drop into the atmosphere and are destroyed. Even manned vessels that have returned to Earth are not used again. All this means that valuable structures representing many millions of dollars are constantly being destroyed or retired from use.

Not so the space shuttle.

The booster will lift itself and the orbiter into the thin upper atmosphere. At a height of nearly forty miles, the orbiter will fire its engines and break away from the booster. The orbiter, having a head start in the booster's own motion, and with very little atmosphere to offer resistance, will be able to go into an orbit that will take it to the space station by the use of very little fuel.

The booster will cruise back to Earth, making use of its jet engine and its aerodynamic properties to take full advantage of the atmosphere. On landing, it will be gone over, refurbished, and made ready for refueling and for repeating its task. There is no reason to suppose that with good breaks and proper maintenance, a booster might not be used many times over.

This represents an important saving. If a single booster is used ten times, it represents a much smaller expense than ten boosters each used one time.

The orbiter will be carrying men or supplies to the space station; objects needed for repairs; new instruments. It will carry back other men or objects, coasting down the gravity hill to Earth's surface again. It, too, can be used over and over again.

The space shuttle can also be used to service and repair ordinary satellites—the weather satellites that automatically check Earth's cloud cover, the navigational satellites that can be used as guides by oceangoing ships on Earth, the communications satellites that will make it possible for all parts of the Earth to be continually in touch with each other.

These satellites will, as a result, be longer-lasting, more efficient, and can be made much more complex and useful. They will free the scientists on the space station from having to do the more mechanical work of observation that automatic devices can do well enough.

The communications satellites, if properly serviced, may be most important of all. Since they will probably be using laser beams eventually, there will be much more room for different wavelength channels for television, radio, and telephone. Already in 1971, "Intelsat IV" is in orbit, with an average capacity for 5000 voice circuits and 12 TV channels.

By 1980, much more elaborate communications satellites will be in orbit and the whole planet may be intimately linked. Perhaps this will make it more possible to develop friendships between peoples, talk out quarrels instead of fighting them out, reach compromises instead of impasses. If the space program ends up doing this, isn't the expense worth it for that alone?

The space shuttle should make it possible to expand the space stations. By the 1980s, stations that are larger, more complex, more efficient, and much more livable will be in orbit.

The advanced space station not only will be a base for

207

observing the Earth, Sun, and Universe, it will become a laboratory for experimentation, making use of the vacuum of space, of the hard radiation from the Sun, of gravity-free conditions; all of which are difficult or impossible to duplicate on the Earth's surface.

The space station may even be a factory that will produce devices impossible to make on Earth. Metals and other substances can be prepared without contamination by gas molecules. Welds can be made more perfectly in the absence of air of any kind. Very thin films could be applied more evenly. Anything that would require a vacuum would have a better one in space without effort than can possibly be produced on Earth except with a great deal of effort.

The absence of a gravitational effect on the space station would make it possible to grow crystals more perfectly, make mixes more efficiently, produce better casts, blow thinner bubbles.

For delicacy and precision, the mark "Made on Space Station" may represent the ultimate.

Most of all, the space station will be the testing ground for life away from Earth. Mankind will learn how to become independent of the planet of his birth. The space station will help us learn that gently. Earth will still be there close by; always ready to reach out and help; always ready to receive back. Then through practice, through trial, through error sometimes, mankind will learn to manipulate the environment of space more efficiently.

As he does this, he will be ready to probe out to the Moon more strongly, more surely, and with much more backing.

By the 1980s, we will have "space tugs" more powerful than any space vessel in existence now. They will be nuclear-powered, so that they will have more thrust per weight of fuel and they will be able to carry large payloads to the vicinity of the Moon.

This "workhorse of space" will be able to carry the men and supplies needed to build space stations much farther out in space; stations that circle the Moon rather than the Earth.

From such a Lunar station, the Moon can be surveyed as the Earth can from our own stations. More than that, the Lunar station will serve as a base for reaching and exploring the Moon's surface in a much more wholesale manner than is now possible.

Specially designed space shuttles will maneuver between the Lunar station and the Lunar surface, and with that a Lunar colony will become possible and, very likely, practical. Life on the Moon would actually be less difficult in some ways than life on a station. There would be the substance of the Moon's crust to serve as raw materials. One could dig under the crust for protection; and find water there, perhaps. One would have room to spread out, and some gravity for comfort.

Since the Moon is airless, observations from its surface would be as efficient as those on the station, but it would have unlimited room for work and equipment. Observatories on the far side, away from the Earth, would be free of the interference of Earth-light during the long Lunar night or from the radio noise emanating from our planet.

The Moon itself could be explored and its geological history worked out. Its geology may tell us more of the evolution of the Solar system than our own would, for whatever took place on the Moon in the way of meteoric bombardment, volcanic action, and Solar radiation has not been obscured by the action of wind, water, and life, as it has been on Earth. What we learn about early history on the Moon will tell us about the early history of Earth, of life, of ourselves.

From the Moon, or from the space stations, we will be launching manned spaceships on the long voyage to Mars by the 1990s. Those spaceships will be far more elaborate than the simple vessels that first carried men to the Moon. They will probably be launched in pairs, each with a crew of three at least. If anything happens to one vessel, the other will be available for repair or rescue.

With the experience gained on the surface of the Moon, it will be that much simpler to operate on the surface of Mars. It is impossible to predict what we will find on the Martian surface, but if even the simplest life forms exist there, the information that will give us, and the help it will offer in understanding the basics of life, may be of incalculable importance.

And, of course, unmanned vessels can be launched in far greater numbers than manned vessels can and for far greater distances. By the time we are ready to send men to Mars, instrument packages will be sailing through space toward every one of the planets of the Solar system; to the asteroids, to the giant planets far beyond, each with their complex satellite system; even to distant Pluto, which it will take many years to reach. There will be probes sent into comets, also; and, in the other direction, to Venus and Mercury, and into an orbit that will skim the Sun as closely as can be managed.

Transportation here on Earth will be affected by all this. Knowledge always laps over its boundaries and it is impossible to advance in one direction without affecting all others. The computer systems used to control and direct space flight may well be used to control the increasingly complex traffic in the air and on the ground here on Earth.

Aircraft may become hybridized; passing through the atmosphere into space, and then back into the atmosphere, in carrying men and cargo from one point on Earth's surface to another.

The vehicles developed for use on the Moon's surface may find applications here and there on the Earth's surface as well and, in the end, a transportation center on Earth in 1995 may turn out to be a combination space flight–air flight nerve center in which some vessels leave for a space station, some for the Moon—and some for other parts of the Earth.

If we can only tackle space exploration with a sure and firm hand, with courage and imagination, and tackle our other more Earthbound problems in the same way, controlling our populations and preserving our environment through increasing our knowledge and wisdom, our children will see a world far different from our own—and far better.

AFTERWORD

I might as well say, by the way, that it is more important in articles dealing with the future, to reach the young people than it is to reach the adults. Adults are, in a way, finished; this is their world and they are impatient with any other.

This is not the world of the young, however. The world

of a quarter century hence is theirs; it is when they will be making the decisions, if they survive. And, in particular, if civilization survives into the twenty-first century and if, under a saner society, we once again venture out into space, it is they who will see it done.

22 COLONIZING THE HEAVENS

FOREWORD

I have been supporting the notion of a Lunar colony for years, as the preceding article would indicate. In 1974, when Gerard O'Neill of Princeton University began to publicize his notions of space colonies as superior to the colonization of world surfaces, I was initially skeptical. I found it difficult to abandon my own set way of thought.

The more I considered the matter, however, the more attractive I found O'Neill's notions, and I became a convert. When Saturday Review asked me, in early 1975, if I had some topic I wanted to discuss, I eagerly suggested that of space colonies. They were willing, and the following essay is the result.

The population bomb ticks on steadily—

We are 4 billion now, in 1975. Barring catastrophes, we shall be 5 billion in 1986, and 6 billion in 1995, and 7 billion in 2002, and 8 billion in—

What do we do with all of ourselves when already,

with our puny 4 billion, we find that the effort to feed and power the population is destroying the planet that feeds and powers us? We must reduce the birthrate and lower the population but that will take time. What do we do meanwhile?

One answer is that we do as we have done before. We must take up the trek again and move on to new lands. Since there are no new lands on Earth worth the taking, we must move to new worlds and colonize the heavens.

No, not the Moon. Professor Gerard O'Neill of the Physics Department of Princeton University suggests two other places to begin with—places as far from Earth as the Moon is, but not the Moon.

Imagine the Moon at zenith, exactly overhead. Trace a line due eastward from the Moon down to the horizon. Two-thirds of the way along that line, one-third of the way up from the horizon, is one of those places. Trace another line westward from the Moon down to the horizon. Two-thirds of the way along that line, one-third of the way up from the horizon, is another of those places.

Put an object in either place and it will form an equilateral triangle with Moon and Earth. It is 380,000 kilometers from Earth to Moon. It is also 380,000 kilometers from Earth to that place, and from Moon to that place.

What is so special about those places? Back in 1772, the astronomer Joseph Louis Lagrange showed that in those places any object remained stationary with respect to the Moon. As the Moon moved about the Earth, any object in either of those places would also move about the Earth in such a way as to keep perfect step with the Moon. The competing gravities of Earth and Moon would keep it where it was. If anything happened to push it out of place it would promptly move back, wobbling back and forth a bit ("librating") as it did so. The two places are called "Lagrangian points" or "libration points."

Lagrange discovered five such places altogether, but three of them are of no importance since they don't represent stable conditions. An object in those three places, once pushed out of place, would continue to drift outward and would never return. The two places in which an object remains are called "L4" and "L5." L4 is the one that lies toward the eastern horizon, and L5 the one that lies toward the western.

Professor O'Neill wants to take advantage of that gravitational lock and suggests the building of space colonies there; colonies that would become permanent parts of the Earth-Moon system.

The vision is of long cylinders designed to hold human beings plus a complex life-support system; facilities for growing food, maintaining atmospheres, recycling wastes, and so on.

Such concepts have been used in science fiction. The most memorable example is Robert A. Heinlein's story "Universe," published in 1941, in which a large ship, supporting thousands of men through indefinite numbers of generations, is making its slow way to the stars. The men aboard have forgotten the original purpose and consider the ship to be the entire Universe (hence the title). A lineal descendant of the story, translated to television, was the recent ill-fated series "Starlost."

In science fiction, though, such enormous self-contained ships are *ships*, thickly spaced with decks, utterly enclosed with walls—the equivalent of many-layered caverns.

O'Neill's vision is of another kind. He sees hollow cylinders with human beings living in the inner surface; a surface which is designed and contoured into a familiar world with all the accouterments and accompaniments of Earth.

The cylinder would be composed of long, alternating strips of opaque and transparent material; aluminum and tough plastic. Sunshine, reflected by long mirrors, would enter and illuminate the cylinder and turn what would otherwise be a cave into a daylit world. The entry of light could be controlled by mirror-shifting to allow for alternating day and night.

The inner surface of the opaque portions of the cylinder would be spread with soil and on it agriculture could be practiced and, eventually, small animals kept. All the artificial works of man—his buildings and machines—would be there, too.

What makes this concept plausible and lifts the vision out of the realm of science fiction is the careful manner in which O'Neill has analyzed the masses of material necessary, the details of design, the thicknesses and strengths of materials required, the manner of lifting and assembly, and the costs of it all. The conclusion is that the establishment of such space colonies is possible and even practical in terms of present-day technology.

It would be expensive, of course, and would require an

input equivalent to that spent on the Apollo program to start the process going, but O'Neill demonstrates clearly that the expense would decline rapidly after that.

As the colonies increase in number, they can be expected to grow larger and more elaborate, too. O'Neill conceives the first space colonies (Model 1) to be only as large as is required to be workable—two spinning cylinders, each 1 kilometer long and 100 meters wide, supporting a total of 10,000 people.

The two cylinders, each spinning about its long axis, would turn in opposite directions. When they are held together this would mean the total system would have virtually no spin and the cylinders could be designed in such a way as to have one end of the structure point constantly toward the Sun in the course of the orbit about the Earth.

It is from the Sun that the colony will be obtaining its energy—a copious, endless, easily handled, nonpolluting form of energy. It will be used to smelt the ores, power the factories, grow the food, recycle the wastes. It will serve to start the cylinders spinning and increase the rate of spin to the point where there will be a centrifugal effect sufficient to hold everything within to all parts of the inner surface with the apparent pull of normal gravity. For a cylinder 100 meters wide this will require a spin of three revolutions a minute.

O'Neill envisions larger cylinder-pairs, too, and has calculated the requirements for some (Model 4) as large as 32 kilometers long and 3.2 kilometers wide, spinning once in two minutes. Each cylinder of a pair like that would be as wide as Manhattan and half again as long and, with a total inner surface ten times as great as that of Manhattan, could support up to 20 million people if exploited to the full, though 5 to 10 million might be a more comfortable population.

With so great a width, there would be a sufficient depth of air within to allow a blue sky and to support clouds. In a Model 4 colony, the end caps of the cylinders could be modeled into mountainous territory—full-sized mountains, not just bas-reliefs.

But where is all the material to come from for the construction of these space colonies? Our groaning planet, sagging under its weight of humanity, with its supply of key resources sputtering and giving out, couldn't possibly afford to give up the colossal quantities of supplies needed for it all. (Over half a million tons of construction is needed for each Model 1; probably a thousand times as much for a Model 4.)

But Earth is lucky, for virtually none of the material need come from our planet. As it happens, we are supplied with a Moon, an empty and dead world that is one-eightieth the size of the Earth. It is close enough for us to reach—we have reached it over and over already—and it is free to be used as a quarry.

Lunar material will yield the aluminum, glass, concrete, and other substances needed to construct the colony. Lunar soil will be spread over the interior surface and in it agriculture will be practiced. Not only is all that material present on the Moon in virtually unlimited quantities, but lifting it off the Moon against that world's weak gravity would require only one-twentieth the effort that would have been required to lift it off Earth. All the smelting and other chemical work would, of course, be done in space.

The Lunar material is not perfectly adapted to human needs. It is low in volatile materials and the most serious lack is hydrogen (an essential component of water, for instance).

O'Neill calculates that to set up a Model 1 colony would require some 5400 metric tons of liquid hydrogen, and that would have to come from Earth. Fortunately, Earth can spare it. We can get it from sea water and there is an embarrassing oversupply of sea water on Earth. We live in comfort only because so much of the Earth's water supply is tied up in the ice caps of Greenland and Antarctica. If these ever melt, the sea level will rise 200 feet and drown our population-packed coastal areas. To extract hydrogen from a little of our oversupply and give it away will do us no harm.

As colonies multiply, of course, the quantity of hydrogen we would have to give up could become a little painful. Once space colonization swings into high, however, hydrogen and other volatile elements in which the Moon is a bit short can be obtained from farther out. They can come from some of the asteroids, or from the occasional comet that blunders past the Earth-Moon system on its way to wheel about the Sun.

The first space colony would be by far the most expensive even if it were small, for we would have to supply not only the advanced equipment, the machinery, the various life forms, the basic food supply and energy, but even some 2 percent of the raw materials. After that, there would be leapfrogging. Each space colony would help to build up the next, while the facilities for mining, smelting, shipping, and constructing would be ever-improving. In the end, new colonies might form with no more trouble than it now takes to put up a new row of houses in the suburbs.

If all were to go optimally, O'Neill thinks that the first space colony could be floating in space by the late 1980s and that several hundred more elaborate colonies would be there by the mid-twenty-first century.

These would be comfortable worlds; not, like Earth, taken as found, but carefully designed to meet human needs. The temperature and weather would be controlled, energy would be free and nonpolluting, weeds, vermin, and pathogenic bacteria would be left back on Earth.

Dangers? Difficulties? Yes, some.

The possibility of a meteor strike exists, but that is not very high. The space of the Earth-Moon system is full of meteoric dust, which is not likely to be bothersome, and pinhead meteors may pit the aluminum and craze the plastic, but that would be a minor annoyance. A meteorite large enough to seriously damage a colony is so rare that the time between strikes could be counted in the millions of years per colony. As the colonies grow more numerous the chances that *one* will be hit increases—but mankind can live with that. We now live with the knowledge that there is a finite chance that at any moment a large meteorite, or a major earthquake, may strike and demolish a city on Earth.

Energetic solar radiation is dangerous but would not be a problem in a cylinder protected by aluminum, plastic, and soil. Cosmic rays are much more serious. They are ever-present and ever-dangerous and very penetrating. There is some question whether O'Neill's original design offered sufficient protection. At the most recent scientific conference held on the subject (at Princeton on May 7–9, 1975), this was among the subjects discussed.

Then, too, the centrifugal effect of the cylinder spin does not perfectly duplicate Earth's gravitation. On Earth, the gravitational pull is not perceptibly altered as we rise from the surface. Inside a spinning cylinder, the effect weakens rapidly as one rises from the inner surface, falling to zero at the long axis.

Is a fluctuating gravitational effect dangerous to the human body in the long run? We have no way of knowing as yet, but if not, a gravitational pull that lessens with height can have its advantages.

The small distances on the space colonies will make it unnecessary to use high-energy systems for transportation. Bicycles would be ideal for the ground and, with the lowering gravity, gliders would be perfect for air transport—and amusement.

Mountain climbing on the larger colonies would have comforts unknown on Earth. As one climbed higher, the downward pull would weaken and it would become easier to climb farther, and, of course, the air would grow neither thinner nor colder. In carefully enclosed areas on the mountaintops, men could fly by their own muscle power when outfitted with plastic wings on light frames. Shades of Icarus!

As the space colonies increased in number, the room available for human beings would increase, too, and at an exponential rate. Within a century, at the most optimistic estimate, there could be room for a billion people on the space colonies, and by 2150, perhaps, there would be more people in space than on Earth.

This does not obviate the need to lower our birthrate in the long run, for if human beings continue to multiply at their present rate, the total mass of flesh and blood would equal the total mass of the known Universe in 6700 years. Long before that the building of space colonies would not be able to keep up under any conceivable conditions.

The colonies could act as a safety valve, however, that would give humanity a somewhat longer time to accomplish the turn-about without absolute disaster.

It might be that finally, when a stable population is attained (or at the very worst, one that is growing only as fast as can be handled by additional colonies), the Earth itself will be only thinly populated. It will, perhaps, be devoted, then, to carefully preserved wilderness and park areas. It may serve as a monument to man's origins and to the prehuman ecology, and it would be supported largely by tourism.

Tourism would exist among the multiplying colonies (which would eventually expand out of the Lagrangian points and take up other and somewhat more difficult-to-handle orbits). Since each colony would have no intrinsic gravitational field to speak of and all are likely to be at about the same distance from the Sun, travel from one to another would consume surprisingly little energy. And since each colony is likely to be unique in its way, the result would be worth the effort.

After all, almost as important as the basic fact that room would be found for humanity is the additional fact that it would be found in thousands of different, isolated, and culturally independent places. Each colony would have its own way of life and some might be quite a distance off the norm. Among the offbeat colonies, we could imagine puritanical ones and hedonistic ones; libertarian ones and authoritarian ones; Orthodox Jewish ones and Hard-shell Baptist ones.

You could choose where you wanted to go and if you were born on one you might choose to try another—or at least visit one. Human culture would explode in variety with each colony having its own styles in clothing, music, art, literature. The options for creativity in general and for scientific advance in particular would be unbounded.

Yet though the advantages of space colonies can be drawn up in a thoroughly lyrical fashion, one has to admit that the chances of the program being started are, perhaps, not bright.

These are difficult times. The Apollo program, by its very success, seems to have taken the shine off space ventures; the economy teeters; there is widespread disillusionment with glowing dreams—particularly in the United States (which would have to supply the major portion of what would have to be a global effort) caught in the wake of the Vietnam failure.

But perhaps mankind will answer to the lure of immediate and high-visibility profit.

With that in mind, perhaps, O'Neill is carefully working out the details of an ancillary idea: the practical economics of establishing a structure designed to be a "Satellite Solar Power Station" (SSPS), one that will absorb Sunlight and convert it into microwave energy which can be beamed to Earth for conversion into direct electrical current.

Earth could, in this way, be supplied with copious, nearly pollution-free energy. The amount of land that would have to be devoted to microwave reception would be, by O'Neill's calculations, as little as 5 percent that required for direct solar-energy reception, because the inefficient part of the operation would be kept out in space.

The development of power stations in space are as practical as the development of space colonies, and the benefits of the former to mankind are sure to be seen as more immediately desirable by the public generally.

Yet once power stations are built in space the expertise and facilities used for that can also be used to build other industrial systems—and space colonies, too. The basic investments having been made, the additional expense will be almost trivial. Why not go on, then?

It is the energy crisis, therefore, that may be offering us the opportunity.

It is Solar energy by way of space that may serve as the bribe.

And if the opportunity is seized and the bribe is accepted, space colonies could follow almost inevitably and with that *might* come the salvation of humanity and its entry into a new and larger scene with overall changes as momentous as those that followed the discovery of fire.

AFTERWORD

This article received a spate of comment from the readers, almost all unfavorable, and almost none sensible. I managed to get a short letter of rebuttal published in Saturday Review, *five months after the article appeared. It went as follows:*

I have received a number of letters concerning my article "Colonizing the Heavens."

Some call it fiction. (Real nonsense, I suppose, like reaching the Moon.)

Some say I am trying to subvert the doctrine of Zero Population Growth. (As though it weren't possible to try to colonize space and stop the population growth, too. They are not mutually exclusive.)

Some say it is too expensive. (Not if the world stops supporting military machines.)

Some say that nobody wants an engineered environment. (Nobody? How many people are living in caves these days?)

Some say that nobody would ever want to cross space

in three days to live in a space colony. (*This from people whose ancestors two or three generations back probably crossed the Atlantic in steerage, or crossed the western desert in covered wagons.*)

Some say that Third World people would never go. (*Sure. Only aristocrats fled to the New World. All the tired, the poor, the huddled masses yearning to breathe free never came, did they?*)

Some say let's solve our problems on Earth before we try to colonize space. (*Someone said that to the Pilgrims. Come on, they said, let's solve our problems right here in Europe.*)

Oh, well.

The letter didn't quite ease my irritation. I proceeded to write an article called "The Nightfall Effect," which is included in my essay collection The Planet That Wasn't (*Doubleday, 1976*).

23 THE MOON AS THRESHOLD

FOREWORD

This article originated as a speech. That is, I was asked by my esteemed publishers, Doubleday & Company, Inc., to give one of the Frank Nelson Doubleday Lectures at the Smithsonian Institution in Washington, D.C. Five such lectures had been given in 1973, five more in 1974, and mine was to be the last of four lectures to be given in 1975.

For the purpose I had to prepare a written manuscript, since the lecture was to be published. (All fourteen lectures were put together in The Frontiers of Knowledge, published by Doubleday in 1975.)

When I gave the lecture in Washington, on April 17, 1975, then, the situation had two unprecedented aspects.

First, I had a written manuscript before me where, ordinarily, I do not even use notes. That was easily handled, though. I put the manuscript down before me upside down and never referred to it.

Second, I was in a tuxedo—the first and only talk I have given in formal dress. Fortunately, the entire audi-

ence was in formal dress as well, so I felt that the misery was shared.

This article was deliberately intended as a broad and overall look at the future of space exploration, as it might be. It therefore incorporates some of the material found in a couple of the earlier essays in this book. However, I wanted to leave the logical development of thought intact here, so I made no attempt to remove the overlap.

Then, too, an earlier attempt at dealing with the subject matter of this article appeared as "The Universe and the Future" in my essay collection Is Anyone There? (Doubleday, 1976) and there is some overlap there. too. An additional decade of thought has, however, sharpened my line of argument, I think, and I wanted the sharpened argument in print.

The twentieth century has seen mankind complete its conquest of the Earth.

On April 6, 1909, the American explorer Robert Edwin Peary reached the North Pole. That meant human beings could range as far north on the face of the globe as they thought it advisable to.

On December 16, 1911, the Norwegian explorer Roald E. G. Amundsen reached the South Pole and extended man's range to the southernmost extreme.

On May 29, 1953, the New Zealand explorer Edmund Percival Hillary and his Sherpa colleague, Tenzing Norkay, set foot upon the top of Mount Everest, that point on Earth farthest above sea level, and mankind's range was extended to the maximum altitude of the solid surface of the planet.

On January 14, 1960, the Frenchman Jacques Piccard and the American Don Walsh descended by bathyscaphe to the bottom of the Marianas Trench in the western Pacific Ocean, and mankind's range was extended to the deepest abyss to which Earth's solid surface plunged.

If there remained any point on Earth's surface, north or south, high or low, that had not felt a human footstep, it was only because no one had yet taken sufficient trouble to reach that point.

But the most remarkable extension of the human range witnessed by the twentieth century has been upward, away from the surface of the Earth. The man-carrying balloon was invented in 1782 and till the very end of the 1800s

223

remained the sole medium by which men could explore the open atmosphere.

On July 2, 1900, however, a human being first rose from the surface of the Earth in *powered* flight when the German inventor Ferdinand von Zeppelin sent up the first dirigible, and on December 17, 1903, the American inventors Wilbur and Orville Wright flew the first heavier-than-air device, the airplane. Within half a century, planes had penetrated the stratosphere and were moving at speeds greater than that of sound. The lowermost twenty-odd miles of the atmosphere were open to man—and that was not the limit either.

On October 4, 1957, the Soviet satellite Sputnik I went into orbit about the Earth at heights where only traces of the atmosphere existed and mankind had made its first move to penetrate space. The first *manned* satellite, carrying the Soviet cosmonaut Yuri Alekseyevich Gagarin, was placed in orbit about the Earth on April 12, 1961, and on July 20, 1969, the American astronaut Neil Armstrong became the first man to set foot upon a world other than the Earth—on our natural satellite, the Moon.

It has been a remarkable leap—from the first powered flight to the landing on the Moon in 69 years and 18 days. That represents a lapse of years less than that of a single normal life; that of my father, for instance, who was born three and a half years before the flight of the first dirigible and who died sixteen days after the first Moon landing.

The question is: where can human beings go next? Can the record of the steady outward extension of humanity's range continue? Can the Moon, having been reached, serve as a threshold to farther regions of space and to conquests far more staggering than anything that has preceded?

Without direct human participation, the exploration of space has extended beyond the Moon, and even much beyond it. Unmanned probes have been sent out into space and have sent back information from successively more distant bodies—from Venus, Mars, Mercury, and Jupiter. As I write, Pioneer 11 is racing toward a rendezvous with Saturn and will, we hope, send back photographs at close range of this farthest of all planets known to the ancients.

The far penetration of human instruments, without human beings themselves involved, does not, however, have the glorious ring of accomplishment that we associate with the mystique of exploration. Plumb lines may be dropped to the ocean depths, but that does not have the same glory of men in bathyscaphes doing the same. Drills have penetrated more deeply into the solid Earth than human beings can go in the foreseeable future, but that scarcely counts as an extension of range.

Never mind mankind's probes; where can *mankind* go beyond the Moon?

It may be that we will be forced to give a disappointing answer here, for though the Lunar landing was a great triumph, the achievement was a deceptively simple one. Earth and Moon are a double world, which occupy what can only be considered neighboring points on any but the very smallest astronomic scale. They are separated by only 380,000 kilometers, a distance equal to only ten times the circumference of the Earth and one that anyone driving a car under average American conditions can exceed in a lifetime of stopping and starting.

Look at it this way. It takes only three days to reach the Moon: and it took seven weeks for Columbus to reach the New World. In reaching the Moon, we have made only the most microscopic dent in the vastness of space. Indeed, we have not really left Earth, since the Moon is as much a slave to Earth's gravitational influence as an apple on a tree is—something Newton perceived three centuries ago.

In going beyond the Moon, we must take our first really large step and here the easy victories are over at once.

The nearest sizable body, other than the Moon, is the planet Venus. Even when it is at its closest to Earth, it is 40,000,000 kilometers away, and that is 105 times the distance of the Moon.

We don't expect a space vessel to move straight across the gap between the planetary orbits. The most economical route for the space vessel to follow is an elliptical orbit of its own that begins at Earth and intersects the orbit of Venus just as that planet is approaching the intersection point (which makes for an intricately calculated trajectory). The space vessel must travel a total distance far greater than the minimum separation of the two planets, and the voyage will take, at best, four months.

Already men have remained in space as long as three months, but that was in Skylab, in Earth's immediate neighborhood, where rescue at short notice was possible. To spend one month longer than that, with every moment taking you farther from home, is a psychological hazard indeed.

Worse yet, having arrived in the neighborhood of Venus, there would be no chance of a landing. That planet, with its thick, hot atmosphere, has a surface temperature of 470° C. everywhere, on the night side as well as the day side. Any exploration of the surface would have to be carried out by unmanned probes launched by the space vessel, which would itself remain in orbit about Venus and would then have to launch itself into another four-month journey back to Earth.

Since exploration of Venus's surface would have to be carried out by unmanned probe, that probe might as well travel all the way from Earth. The benefits achieved by having the probe launched from, and signals received by, a manned mother ship would scarcely justify the traumatic experience of over eight continuous months in space.

Mercury, the planet nearest to the Sun, is farther from us than Venus is, being never closer to us than 80,000,000 kilometers. The elliptical orbit that would take astronauts there is not enormously longer than that to Venus, and astronauts can at least land on Mercury.

Mercury has only a trace atmosphere incapable of conserving heat. In the course of its slow rotation (57 days), the Mercurian surface can heat up in spots to temperatures higher than that of Venus, but in the course of the Mercurian night, the temperature of the surface would drop rapidly and present no problem. Astronauts could explore a given area for a month between sunset and sunrise.

Although the total elapsed time represented by a flight to Mercury and back would be greater than that to Venus and back, the former would be broken midway by a planetary landing, a chance to uncramp from the confined quarters of the space vessel, and that would make the trip to Mercury less psychologically difficult than the one to Venus.

The flight to Mercury would, however, carry the astronauts and their ships to some 65,000,000 kilometers from the Sun, even when that small planet is at its greatest distance from the Solar furnace. Solar radiation would be over four times as concentrated at that distance as it is in the neighborhood of the Earth. For what might be gained in a long, manned voyage to Mercury, the price paid in risking the effects of the greater radiation may prove too high.

Since voyages in the direction of the Sun offer no suitable target, what about voyages away from the Sun?

The nearest planet to Earth in the other direction is, of course, Mars. It is, at its closest, some 58,000,000 kilometers away, so that while it is farther than Venus, it is closer to us than Mercury is, and traveling Mars-ward means steady progress in the direction of decreasing intensities of Solar radiation. Furthermore, it is a cool world (cooler than Earth) and can be explored in comfort for indefinite periods whether the Sun is in the sky or not. There is no fearsome Sunrise acting as a deadline, as there would be on Mercury.

In almost every respect, Mars is more interesting than the Moon. Mars is the larger world; it has an atmosphere, albeit a thin one; it has some water, albeit very little; it has polar ice caps and seasonal changes; it has an active geology that produces volcanoes; it has the possibility of long-range cyclical changes that may produce a milder Mars, at intervals, with a thicker atmosphere and free liquid water, and consequently the possibility of a native life.

But in one respect, Mars is a more difficult target than the Moon is. It is 150 times as far away as the Moon is, even at the point of closest approach.

The round trip, to Mars and back, would take over a year at the very least. Even though that will be broken, for a shorter or longer time, by a landing on a planet which, next to Earth itself, is the most comfortable in the Solar system, the task would surely stretch human endurance to the limit.

And beyond Mars? To reach the larger asteroids, or the asteroids of the giant outer planets, Jupiter, Saturn, Uranus, and Neptune, would take years and even decades. Manned voyages of such lengths do not seem practical at the moment.

In addition to the Moon, then, we are left with only Mars as a sizable target and that only as a borderline possibility. There would also be the smaller bodies that venture within the Martian orbit. These would include the occasional asteroid such as Eros or Icarus, and the occasional comet such as Encke's comet. For completeness, we might also mention the two tiny satellites of Mars, Phobos and Deimos.

All in all, the possibilities beyond the Moon offer us a small and disappointing range of targets. It would seem that in the present state of the art, the Moon not only is the threshold of space but, as far as manned exploration is concerned, is virtually all of it.

Yet surely it is a mistake to limit oneself to some existing state of the art, without considering the possibility of advance. To have done so in 1900 would have made the Moon unreachable to mankind; to have done so in 900 would have made North America unreachable to Europeans.

The difficulty in moving to the worlds beyond the Moon is one of distance, and of the time it takes to cover that distance. Years and decades in space in the kind of vessels we can now build are simply not in the cards.

Can distance be eliminated altogether by removing the necessity of a slow acceleration consistent with what the human body can endure, and, therefore, the achievement of high speeds in a very short interval of time? If we could combine this with the attainment of speeds far in excess of the speed of light, not only might the outer regions of the Solar system be reached in a reasonably short time, but the stars themselves might fall within our grasp.

(As long as the speed of light remains a limit, even relatively near stars require round trips lasting decades, while moderately distant stars require centuries of travel. To cross our galaxy would take some hundreds of thousands of years, and to reach any object outside our galaxy we will find ourselves dealing with journeys of millions and billions of years in length.)

How can we possibly bypass the speed of light, though, when that is the ineluctable limit set by the theory of relativity? Some physicists, however, have postulated particles of matter that have the property of traveling only at speeds *greater* than that of light and show that this, too, is consistent with relativity. Either always slower, or always greater, but never both. The faster-than-light particles are called "tachyons," from a Greek word for "fast."

We might imagine, then, that every subatomic particle making up a ship and its contents might be translated into the corresponding tachyons instantaneously. These would travel at enormous speeds for some desired period of time in the form of "tachyonic ships" carrying a "tachyonic crew." They would at a given moment change back, again instantaneously, into ordinary particles. In less than a second, a ship might, in theory, travel any distance—from end to end of the Universe perhaps.

Tachyons, however, have been only postulated, so far; they have not been detected. Even if they existed, the necessity of changing *every* normal particle into tachyons, and then reversing the change, in virtually perfect simultaneity, would be a most harsh requirement. (If some particles make the change a fraction of a second out of tune with the others, the tachyonic ship and crew may end up spread over light-years of space.) Finally, even if the changes could be made, the task of directing and controlling tachyonic flight would represent a most formidable engineering problem.

All told, while we can speculate about tachyonic flight, it is not reasonable to consider it seriously. It can come about only as a result of a fortunate and an as-yet-looked-for series of scientific breakthroughs, which may very well never happen.

Even if the speed-of-light limit is never surpassed, it might still be possible to eliminate the sensation of time passage. We might imagine a ship traveling through interplanetary and even interstellar space at quite a slow velocity, but with its astronauts in a frozen state of suspended animation, and the ship itself under strictly automatic control. On approaching some destination, the astronauts can be automatically thawed and roused.

This again implies the development of techniques that, at present, we have no right to assume as an inevitable development. Relatively simple life forms can be frozen and revived, and it seems quite likely that living organisms frozen quickly enough and brought to low enough temperatures (say, those of liquid air) might remain suspended indefinitely in a state that can be brought back to life.

Nevertheless, the difficulties of freezing a living object as large as a man sufficiently quickly to produce no life-destroying damage in the process are enormous. No warm-blooded creature has as yet been thoroughly frozen and then restored to life, and we cannot be confident that this will ever be successfully brought about. We must therefore allow frozen flight to remain, along with tachyonic flight, a subject for far-out speculation only and not for serious anticipation.

229

There is, however, another way of eliminating the sensation of time passage while allowing the astronauts to retain full consciousness; a method that is known to be possible in the light of modern knowledge and that requires no unforeseen breakthroughs.

The theory of relativity makes it quite clear that as the speed of a vessel increases, relative to the Universe generally, all motion of whatever type slows, meaning that the time rate that is experienced also slows, reaching zero at the speed of light.

If a ship's velocity is raised to near the speed of light, then to the astronauts on board, a voyage that would ordinarily be experienced as having endured centuries, would seem to have endured, to the vastly slowed time sense, only years or even weeks. Under such circumstances, the astronauts by edging closer and closer to the speed of light could cover longer and longer distances in a given time interval and would, in effect, accomplish what a tachyonic ship would.

There are, however, catches to travel by time dilatation as compared to travel by tachyons that removes some of the bloom from the former.

1. If we could turn ordinary subatomic particles to tachyons, that might, in theory, be done virtually instantaneously. To attain near-light velocities, however, there must be acceleration first and then, as the destination nears, deceleration. The rate of neither acceleration nor deceleration can be very great in view of the fragile structure of the human body, so the task of getting up to high speeds and getting down from it is very time-consuming (and very energy-consuming as well). Interstellar travel will therefore always take some irreducible and uncomfortably long time despite the time-dilatation effect.

2. Once a very high velocity is achieved, there is a problem concerning the interstellar medium itself. We consider it a vacuum and it is certainly a better vacuum than anything we can make here on Earth. At very high speeds, however, a ship will pass so many of the thinly spread atoms that exist even in interstellar space, that what seems a vacuum at ordinary speeds will become a highly resisting medium at high speeds, and that may set a practical limit to how fast one can go. That limit may be sufficiently far from the speed of light to reduce the time-dilatation effect considerably.

3. Even if the time and energy problem is dismissed and if we assume that near-light velocities can be achieved without interference from the interstellar medium, the fact remains that time dilatation affects only the speeding astronauts and not the rest of the Universe generally. The space vessel and its crew may indeed be able to reach the Andromeda galaxy in, let us say, a year, explore some part of it for a year, and return in another year—but on Earth, lacking the time-dilatation effect, four and a half million years would have passed. (This sort of thing does not happen in the case of tachyonic travel.)

It is doubtful if astronauts would volunteer for a space voyage if they knew that they would never return to any Earth they could recognize. And even if some would indeed be willing to wash their hands of Earth forever (perhaps with a muttered "Good riddance!") it is considerably more doubtful that Earth would be willing to invest in an exploring mission of which no human being then alive (and possibly no human being who would ever live) would see the results. The question "What's in it for me?" in connection with the expense and effort of the voyage would have as its answer "Nothing!" and so it would not be done. It might be, then, that only the very nearest stars would be within reach, no matter how high a velocity could theoretically be reached and no matter how slow time could be made to travel.

There seems to be only the alternative of accepting the limitations of both time and distance and of making the best of it. One can point a ship at the stars, let it gain some moderate velocity, and then let it coast indefinitely. To reach even the nearer stars, under such circumstances, might take thousands of years as judged by the astronauts themselves as well as by the stay-at-homes on Earth. Generations of astronauts would have to be born, live out their lives, and die, in the course of so mighty a voyage.

Naturally, this could not be done on board any spaceship of the kind with which we are now familiar. There is no life-support system we can now build that would last for years, let alone millennia, and it isn't conceivable that astronauts would be willing to pass their lives in constricted quarters and have children who would then have to pass *their* entire lives in those same constricted quarters.

It would be necessary to have a ship large enough to represent a world—a small one, perhaps, but one sufficiently elaborate for each member of the crew to find his immediate surroundings world-like. The dimensions of the ship would have to be measured in tens of kilometers and the population on board would have to number in the hundreds of thousands.

Let us call such large vessels, capable of supporting a large population indefinitely by on-board agriculture, animal husbandry, and industry, a "starship."

The concept of the starship requires no unforeseen breakthroughs and seems to possess no hidden technological catches—but there are psychological problems:

1. Merely building a starship would require some tens of billions of present-day dollars, some tens of billions of tons of materials of all kinds, including a vast quantity of energy. Even if such a project is feasible from the technological standpoint, would the population of Earth be willing to devote the effort and the resources for the purpose?

2. This is especially so since, as in the case of a time-dilatation ship, there is no chance of a starship returning to Earth except after huge lapses of time. Again, the population of Earth would scarcely be willing to invest so much when there will be no visible return of any kind in the foreseeable future.

It would seem, then, that even looking hopefully into a future of advanced technology, we may not make many advances in the exploration of space. It almost seems that no matter in what direction we look we end with the Moon being both beginning and end; no human exploration seems possible beyond the Moon—now or ever.

But can this be so? The thought of so restricting a limitation is abhorrent to any romantic (and all explorers must be romantics) and we must therefore look further and search avidly for any loopholes that might brighten the bleak picture I have drawn.

Through the discussion so far, I have made the assumption that the home base for space exploration will be the Earth; that it will be from Earth that our explorations will start. Might it not be, though, that if we can broaden our base, the situation with respect to the further exploration of space might change—and for the better? After all, it could hardly change for the worse.

And surely our base will broaden. If our civilization

232

survives* it will inevitably broaden. However narrow our capacities for exploring space may be, they are broad enough to include the Moon. We have already reached the Moon and returned safely six different times. If we develop the space shuttle and establish space stations more advanced and versatile than Skylab, we can inititate flights to the Moon that will cost far less than those of the Apollo program and be capable of accomplishing more. Given a reasonable advance in technology, we can establish a permanent, ecologically independent colony on the Moon.

The Moon seems a harsh and forbidding world by Earth standards, but most of those factors that make it seem forbidding are properties of the surface only. It is the surface that is subjected to a two-week siege of Sunlight that brings the temperature, in spots, to the boiling point of water—followed by a two-week absence of Sun that lowers the temperature halfway to absolute zero. It is the surface that is subjected to meteorite bombardment, to the harsh radiation of the Sun, and to the cosmic radiation from points beyond. It is the surface that lacks air and water, and could not hold on to them (thanks to a surface gravity only one-sixth that of Earth) even if air and water were somehow supplied and placed upon the surface.

Once caverns are excavated beneath that surface, though, and several meters of Lunar soil are placed between its deadly properties and humanity, a comfortable world could be built. Beneath the surface, temperature would be equable and uniform and there would be neither day nor night, neither summer nor winter. With artificial lighting, time could be organized for human convenience. There would be no danger from meteorites† or from radiation either.

To begin with, of course, there would have to be a large capital investment from Earth. Human beings, plants and animals, machinery, energy, would all have to be supplied by Earth at first. Little by little, though, the Lunar colon-

---

* This is an enormous "if" which I by no means wish to minimize. As the crisis brought on by heedless overpopulation and criminal waste of resources gathers about the collective head of humanity, it is only too likely that we may destroy our civilization in the course of the next thirty years—and if so, this essay will merely be a glimpse into the might-have-been and nothing more.

† A really large meteorite would be deadly, of course, but these are excessively rare—and they would be deadly on Earth, too, after all.

ists would begin to draw upon the Moon itself for resources.

Energy would be easily available on the Moon, for instance. Any given spot on the Moon experiences Sunlight for two weeks at a time, with no interference ever from clouds or dust—and points on the Lunar equator experience the maximum quantity. Solar cells could be spread out over almost unlimited areas. There is no weather to interfere, no vandals to fear, no native life forms to displace, no native ecology to upset.

The energy so obtained could be used to power the chemical processes that will release from the molecules of the Lunar soil the various elements which can be combined to form other molecules. The elements themselves, the molecular compounds, whether native or synthetic, would be used in construction, in the building of machinery, in the million and one uses already worked out on Earth.

Energy, in the form of an electric current, could split water into hydrogen and oxygen. The oxygen could be used in building an atmosphere, the hydrogen in chemical syntheses. Unicellular organisms, growing rapidly under the encouragement of human-expired carbon dioxide, artificial light, chemical fertilizers, properly treated human waste, could form protein at breakneck speed with almost no waste. Some of them could renew the atmosphere and all would remove waste and serve as a food supply.

Eventually some forms of animal life could be introduced and multicellular plant life. A reasonably normal human dietary might be established.

A major flaw in this picture of a flourishing Lunar colony has been revealed as a result of the manned exploration of the satellite. The rocks brought back to Earth show that the Moon's crust is low in the content of the more volatile elements; those that form compounds that are low-melting. Presumably, the Moon went through more or less extended periods at elevated temperatures and lost volatiles by vaporization.

In particular, water is absent. Judging by the nature of the Lunar rocks that have been studied, it would seem that the Lunar crust is everywhere thoroughly dry.

This may turn out to be an overly pessimistic conclusion, but even if we allow a totally dry Moon, a Lunar colony need not be ruled out. Earth has, if anything, an oversupply of water, for if a slight rise in temperature were to melt the planetary ice caps, the coastal rims of the continents would be drowned, producing unimaginable disaster. It would be no great sacrifice to offer the Lunar colony some tens of thousands of tons of water; with additional hydrogen, perhaps, for other uses.

Since the Lunar colony would carefully recycle everything as efficiently as possible (as, otherwise, survival would be questionable), not much in the way of new material would have to be introduced. Small capital investments of water, for instance, would last a long time. Additional supplies would be required more to sustain colony growth, perhaps, than to replace cycling losses.

To be sure, it may be that Earth, over-aware of its own needs, would choose to grant only grudgingly and sparingly the water required by the Lunar colony. It may also be that the Lunar colony, oversensitive to its own dependence on the home world, would seek some water source other than Earth.

It might seem, at first, that there is no other practical source within reach for volatiles (not only for water but for compounds of such elements as carbon and nitrogen in which the Moon is relatively lacking and which are particularly important to life).

Venus possesses volatiles but these are present in its atmosphere in gaseous form and would therefore be difficult to gather in quantity. Mars has sizable ice caps containing frozen water and frozen carbon dioxide, but Mars offers the next possible base for the expansion of humanity and there would be serious ethical questions as to whether that planet's limited store of volatiles ought to be rifled.

But where else can the Lunar colonists turn? It is in the outer Solar system, where the intensity of Solar radiation has always been low, that the real supply of volatiles is to be found. Some of the larger satellites may possess significant quantities. Ganymede and Callisto, which circle Jupiter, and Titan, which circles Saturn, are thought to be rich in volatiles.

The outer Solar system is at a great distance, however, and the Lunar colonists, at least in the early years of their existence, would surely feel the need of something closer.

Fortunately, there is an alternative. Not everything that exists in the outer Solar system remains there permanently. There are some objects in the Solar system that have highly elongated orbits—the comets. At the far end of their orbits, they are in the outer Solar system, far beyond even the farthest planet in some cases. At the other end of their orbits, they pass through the inner Solar system.

The comets, originating (it is thought) in the far reaches of space, a light-year or more from the Sun, consist, to begin with, of "ices"; that is, of frozen volatiles. (Observations of comet Kahoutek from Skylab in 1974 offered the final confirmation of this.)

If comets remained in the far-off cloud in which they formed, they would remain frozen and intact indefinitely. Every once in a while, however, a comet slows in its orbit because of the pull of some distant star and drops in toward the Sun.

Comets lose some of their volatiles at each entry into the inner Solar system and each whirl about the Sun. Those volatiles boil off to form a foggy "coma," which is then driven outward by the Solar wind into a long filmy tail. Comets which take up relatively short orbits (because of the gravitational perturbations of the planets they pass) approach the Sun at frequent intervals, losing their volatiles in time and leaving behind a rocky core, or nothing more than a thin cloud of dust. There are always comets, however, which have made, as yet, few approaches, and which are still rich in volatiles.

It may be that, in time, the Lunar colonists, having struggled along with what skimpy supplies of volatiles they have squeezed out of a reluctant Earth, will have developed the techniques for trapping such comets.

Moon-based telescopes would detect such comets far out in space, even before they had reached Jupiter's orbit. (To be spotted far off is itself the sign of a large new comet, possibly very rich in volatiles.) The comet's orbit can be plotted and in the months it takes to approach and enter the inner Solar system, the Lunar colonists will have placed a ship at some rendezvous point in space.

A landing will be made on the comet, which may be no more than a few kilometers across the solid core. Rockets appropriately placed on the comet which make use of the substance of the comet itself as the material to be turned into exhaust once heated, or possibly the use of some advanced nuclear drive, would force the comet out of its orbit.

Little by little, the comet's motion will curve in such a way as to bring it slowly closer to the Moon, then into orbit about the Moon, then spiraling down to the Moon's surface. Finally, it might be brought down within the southern lip of a north-polar crater, for instance, where, in the eternal shadow of the crater wall, it will remain permanently frozen.

The whole process will be like that of hooking, maneuvering, and landing a gigantic fish. Such a "beached" comet, with cubic kilometers of volatiles, will easily make the Lunar colonists independent of further supplies for decades. And long before the supply is consumed, another comet may have come into view.

It is not necessary to discuss here the uses and values of a Lunar colony. These are many, for it can serve as a base for research that cannot be performed on Earth; a place where industrial processes can take place in super-Earthly fashion by making use of un-Earthly conditions of low temperature, high vacuum, intense Solar radiation, and so on; a place which will serve as a first attempt at preparing a thoroughly engineered abode for human beings and as a working model for Earth of a low-birthrate, resources-conserving-and-recycling economy by which alone our home planet can survive.

All this will eventually make the Lunar colony hugely profitable to Earth, but never mind that. What I shall discuss here is how the existence of a flourishing, prosperous, vigorously growing Lunar colony would affect the program for the further exploration of space.

The prospects will surely change dramatically once the Lunar colony is a going concern, for the interaction of Lunar colonists and space is bound to be entirely different from that of Earth people and space.

1. Space flight is an exotic matter to the people of Earth, something that would take them away from the world on which they live and on which they have developed over a period of billions of years. Space flight would, on the other hand, be of the very essence of life to the Lunar colonist, whose world was populated as a result of space flight, and which would be growing and prospering only through volatiles obtained by space flight (whether from Earth or elsewhere). Where Earth people might, on the whole, hesitate to venture into space, that would come as second nature to the Lunar colonist.

2. The conditions of space flight represent an extreme change-about to us of Earth. We are accustomed to clinging to the outer surface of a very large world; to a cycling of food, air, and water through so vast a system that one is scarcely aware of it; and to all the accouterments of

such a world—the blue sky and green land, the sound of birds, the smell of flowers, and all the rest. Getting into a spaceship would mean inhabiting the inside of a very small world; the cycling of food, air, and water in so tight a fashion as to prove an ever-present fact forcing itself on the consciousness of the crew. And, of course, there would be none of the pleasant side-effect properties of Earth.

For Lunar colonists, however, the change is not at all extreme. They live, in any case, on the inside of a world which is, to be sure, a fairly large one taken in all, but in which the cavern volume that forms the effective habitat is small. The colonists are accustomed to close cycling and to a thoroughly engineered environment. They have none of the accouterments of Earth.

In short, the Lunar colonists, in undertaking a space flight, move from one spaceship to another very similar but somewhat smaller spaceship.

This does not make the space flights to some specific destination less long or less dangerous, but it does enormously lessen the psychological difficulties. A crew of Lunar colonists could undoubtedly endure the restricted quarters of a spaceship for a year or more far more stoically and efficiently than an Earth crew could.

3. Since the surface gravity of the Moon is only one-sixth that of the Earth, and since there is no atmosphere on the Moon, a spaceship takeoff on the Moon requires far less energy than it does on Earth, and runs no risk of overheating through friction. What is more, most of the destinations for space flight in our Solar system are much more Moon-like in character than Earth-like—gravitationally and in every other way. The Lunar colonist would feel more at home on asteroids and satellites than ever Earthpeople would.

4. The fact that the Lunar colonist will be accustomed to a gravity only one-sixth that of Earth is important. Developing under a low gravity, he is likely to possess slimmer and more delicate bones and muscles and may be more dextrous and delicate in the manipulation of machinery and controls. It may also be that he is less likely to suffer loss of muscle tone and bone calcium under prolonged subjection to zero-gravity conditions, losses that have affected Earthpeople so subjected.†

‡ To be sure, I am assuming here that it will be possible for human beings to be born under low-gravity conditions, and to grow and develop with full physiological functioning. We won't certainly know till we try.

It may well be, then, that Lunar colonists can make long trips under zero-gravity conditions which to Earth-people would be impossible.

Suppose, though, that Lunar colonists are adapted to low-gravity but cannot withstand zero-gravity conditions any better than Earthpeople can. It may then be necessary to produce a gravity substitute in the form of the centrifugal effect by placing a spin on the vessel.

The effect depends for its intensity on the rate of spin and on the distance of an object from the center of rotation. If the Lunar colonists are well adapted to lower gravity, any ship they are on will need a rate of spin only one-sixth as rapid as that of an equivalent ship with Earthmen aboard—thus reducing the engineering problems.

If we take all this into consideration, it may well prove that although Earthmen may reach Mars as a *tour de force* before a Lunar colony is brought into full flourishing existence, it will be only after that colony is mature and only after Lunar colonists, born and bred on the Moon, are launched into space, that trips to Mars and back become a routine affair. It is only with the voyages of the Lunar colonists, facing a year or two in space with equanimity, that the Martian surface will feel the footsteps of human beings in wholesale quantities; that the planet might be explored; and that the first engineering steps may be taken to make it safe for mankind.

Of course, the Lunar colonists might not, themselves, colonize Mars. The Martian surface gravity is two-fifths that of Earth, but still two and a half times that of the Moon. This raises a problem, for one can perhaps travel from an accustomed strong gravitational pull to an unaccustomed weak one with no pain and with easy adaptability, but the reverse can scarcely be easy.

The Lunar colonists, then, may find it uncomfortable to remain very long on the Martian surface. They may have to establish their base on Mars's tiny and nearby satellite, Phobos, and explore and engineer the planet in shifts.*

Once Lunar colonists have done their work, however, people from Earth can be brought to Mars as colonists in modified Lunar ships (larger and of faster spin) manned

* Naturally, Lunar colonists will find it far more uncomfortable to be on Earth's surface and may, in fact, never visit Earth and may feel that to be small loss.

by Lunar colonists. Mars may then form a third world inhabited by human beings.

Once it is reached, Mars may prove far easier to colonize than the Moon would be. The Martian gravity is closer to that of Earth; it has a thin atmosphere, though an unbreatheable one; and it has an Earth-like rotation of just over twenty-four hours. Its greater distance from the Sun keeps the surface temperature cool and minimizes the danger of Solar radiation. (Since it is closer to the asteroid belt, Mars may suffer from a somewhat greater meteorite bombardment intensity, however.)

Most of all, Mars has an ample supply of volatiles of its own. Certainly, the quantity is very small on an Earth scale, but after all, the supply will not have to meet the needs of a planetary ecology, but only those of a limited Earth colony.†

In some ways, the Martian colonists will enjoy the advantages of the Lunar colonists as far as space flight is concerned. Like the Lunar colonists, the Martian colonists will be living in an engineered underground world and will not feel a strong psychological wrench in transferring to a smaller engineered underground world. What's more, the Martian colonists will be closer to the vast unexplored reaches of the outer Solar system.

Still, Mars, with its atmosphere and its greater gravitational field, is a more difficult base from which to launch spaceships, and its colonists will find it harder to adapt to zero gravity. On the whole, then, the Lunar colonists may retain their virtual monopoly on major space flight.

In fact, we might envisage a twenty-first century in which Earth and Mars are occupied by relatively immobile and "landlocked" races of *Homo sapiens,* while the Lunar colonists will launch themselves fearlessly into the vast ocean of space, playing the role, over a greatly enlarged sphere of action, of the Phoenician and Polynesian voyagers of the past.

The voyages of the Lunar colonists will undoubtedly be important even if they land nowhere, for they can test the properties of the space surrounding the major planets and at varying distances from the Sun. Still, they might easily establish bases in the outer Solar system. Except for

---

† I assume here, by the way, that Mars does not, in fact, have a living ecology of its own. It may not, but we can't be sure; in fact, astronomers and biologists hope fervently that it does. And if it does, it may be ethically improper to colonize the planet.

the giant outer planets themselves (which are likely to remain unapproachable by mankind in the foreseeable future), all bodies in the outer Solar system have surface gravities in the Lunar range or less and would be easily occupied by the Lunar colonists.

The larger asteroids are obvious targets for bases where a spaceship on a multi-year flight may stop for "rest and relaxation" while the asteroid, moving in its orbit, carries the ship into a more strategic position for taking up a new portion of its own trajectory.

The same might be said of the satellites of the outer planets, though here difficulties might arise in connection with the magnetic fields of the primaries. The large Galilean satellites of Jupiter are within the enormous Jovian magnetic field and human penetration of that field would raise serious questions as to shielding. Callisto, the farthest of the Galilean satellites, may be relatively safe, and certainly the small outer satellites of Jupiter (eight of these are known at the moment) would be.

We might visualize numerous bases established throughout the Solar system in time, one even as far out as distant Pluto, where the initial purpose of the colony might be to devote itself to the study of the stars in a sky in which the Sun is reduced to nothing more than an extraordinary bright star itself. Pluto, with a surface gravity about that of Mars, may have extremely low temperatures, but, given energy, it is easy to heat low-temperature caverns. While Solar energy would be unattainable on Pluto, by the time it is reached, mankind ought to have useful nuclear fusion reactors and the planet should have a supply of hydrogen to use as fuel.

The Plutonian colonists may represent a third relatively immobile race of *Homo sapiens*, one that would be fearfully isolated‡ and would be utterly out of reach of other human beings except by way of radio communication (itself not an easy task over enormous distance where it would take hours for a message to go and an answer to return, even at the speed of light). There might be an occasional visit by a Lunar ship, however, or perhaps, more likely, by a Callistan or Titanian ship.

And in ships better shielded against radiation than any

---

‡ We must not overestimate the effect of isolation, however. The population of Earth itself is fearfully isolated today since we know of no other inhabited world—not one—and yet the sense of isolation does not overwhelm us.

we can build now, the Lunar colonists may head inward toward the Sun, studying the properties in the neighborhoods of Venus and of our luminary itself, and effecting an actual landing on Mercury.

If all this comes to pass, then, we will find the Moon to be indeed a threshold to space. While merely reaching the Moon does us little good, colonizing the satellite would open the entire Solar system to human colonization and development. The result of that would be to greatly extend the knowledge that will be available to human beings; greatly multiply the variability of human culture; and greatly increase human security in case of disaster, since no longer would all human beings be found on a single world.

If the Moon is the threshold to the exploration and colonization of the entire Solar system, will it help us move beyond the Solar system to the far greater Universe of the stars?

The various advances in technology that might make the exploration of the stars possible—faster-than-light drives, near-light velocities, long-term freezing—have been considered and found wanting with Earth itself as an exploration base. They are still to be found wanting and their uncertainties and flaws are not in the least ameliorated by the fact that mankind is on the Moon, or on any other world of the Solar system. With the entire Solar system triumphantly penetrated, exploited, and colonized, the stars will still remain unreachable, in all likelihood, by such techniques.

But what of the last alternative? What of the construction and use of a starship? Does that become in any way more reasonable and practical because there are colonists on the Moon?

In some ways, yes.

The construction of a starship on the Moon would be very much equivalent to the construction of a cavern extension under the Lunar surface. The Lunar colonists would have the necessary skills and experience and could build a starship more efficiently than Earthpeople could.

What's more, the materials for the starship would be drawn from the Moon itself. Since the Lunar surface is dead and since the Lunar colonists will, for a long time, take up only a small portion of that Lunar surface, there will probably be no sense of loss in starship construction on the Moon.

Earthpeople are bound to feel that every ton of aluminum, steel, concrete, or glass taken from Earth and used for a starship is a ton abstracted from possible use by Earthpeople and a ton whose removal diminishes the strength of an already enfeebled Terrestrial ecology. The Lunar colonists, on the other hand, are likely to shrug off the ton used as a ton that would otherwise lie in the ground uselessly for an indefinite period of time.

Once built, a starship manned by Lunar colonists would offer an environment even more like the Lunar colony itself than an ordinary space vessel would be, and the starship could be launched from the Moon with far less energy expenditure than it could be launched from Earth.

Yet even so, one can imagine disadvantages. For one thing, there remains the difficulty involved in leaving home *forever*. Lunar colonists who might be willing to undertake voyages lasting for years may balk at voyages lasting forever. Worse yet, the Lunar colonists remaining at home may balk at building a ship that will never return. The inhibiting effect of "Goodbye forever!" might well be as effective on the Moon as on the Earth in stalling the starship project.

There is also, perhaps, a physiological difficulty. A starship, we might argue, ought to have some gravitational effect. Even if men adapted to low gravity could endure zero gravity over space flights with durations of several years, it might be that the actual birth of children under zero gravity and their growth and development under zero gravity would prove impossible.

It would be a difficult thing to check, but the testing of birth and infant development in an orbiting satellite might be attempted, and it might turn out to be feasible. Zero-gravity children might develop a whole new set of conditioned reflexes and learned responses that would make it second nature for them to move and to handle objects under zero gravity, so that they will do it as easily and deftly as we work under normal gravity.

Even if that were so, however, a starship under zero gravity would not be advisable. It would make of the human population of the ship permanent prisoners. They might well be unable to step out upon any world larger than a moderately small asteroid. It might be that any perceptible gravitational pull, one of only a pound or two, might be frightening, nauseating—even physiologically unendurable.

Surely, if a starship is a true exploring device, its crew should be able to step out on a reasonably large world, and not needlessly limit their range to bodies of insignificant size only.

A large ship, set spinning at a moerate rate about its long axis, would force objects toward the inner surface of its shell, and away from the axis in all directions. Wherever one stands in such a ship, the direction toward the axis would be up and the direction toward the shell and open space would be down (rather the reverse of the situation on a sizable world).

This, however, might introduce a difficulty.

In any world large enough to possess an appreciable gravity, the gravitational intensity decreases with distance from the center, but that distance is already so large at the surface that it takes a considerable further increase to produce a noticeable decrease in gravity. One would have to rise to a height of 35 kilometers above sea level on Earth (four times the height of Mount Everest) before gravitational intensity drops by 1 percent. It would require a height of 19 kilometers above the average Lunar surface level to cause the Moon's gravitational intensity to drop by 1 percent.

On such worlds, the gravitational intensity is roughly constant, when one is on or near the surface and all the sensations, reflexes, and responses are conditioned to a constant gravitational intensity.

A cylindrical starship, three kilometers across the short axis, let us say, and rotating about its long axis, would produce a sensation of gravitational attraction toward the shell that would decrease linearly as one moves toward the axis, reaching zero at the axis. If one moved fifteen meters away from the outer shell toward the axis, the sense of gravitational pull would drop 1 percent in intensity.

Moving through the ship would surely involve relatively large changes in gravitational effect, which would itself produce physiological, psychological, and engineering difficulties.

Undoubtedly, if we were liberal enough to suppose that the human body can adapt itself to low gravity and even zero gravity, we might go a step further and assume it can adapt itself to variable gravity as well—but it nevertheless adds one more difficulty to the "Goodbye forever" syndrome.

For these several reasons, then, it might be that even with a Lunar colony in successful operation, and with a Solar system fully developed, the exploration of space beyond the Solar system might be stymied.

It may be, though, that we are overlooking another way of expanding the human base for exploration, one that could be perfectly feasible and that might then make exploration beyond the Solar system a completely practical matter.

Suppose that the Lunar material and Lunar expertise were used to build a starship, but that the starship was *not* launched on its endless voyage into space, but was kept in the Solar system.

Gerard K. O'Neill of Princeton University suggested, in 1974, that the equivalent of what I have called starships could be built in the near future and made into orbiting space colonies. In his view, the Lunar colony, which has always been taken for granted as the first step in the expansion of humanity beyond Earth, could well be overleaped. People would land on the Moon, yes, but not to colonize it; only to convert it into an enormous quarry. Material from the Lunar surface could be towed out into space where it could be smelted and converted into steel, aluminum, and glass, then put together as a large container for soil (also from the Moon) together with machinery, plants, animals, people, and liquid hydrogen—all of which would be initially from Earth.

O'Neill visualizes the first space colonies as being placed in the Trojan position with respect to Earth and Moon. The three bodies—Earth, Moon, and space-colony cluster—would form the vertices of an equilateral triangle. The colonies would be 380,000 kilometers from Earth and 380,000 kilometers from the Moon. The colonies would move about the Earth in the Moon's orbit, lagging behind the Moon by 60 degrees, or moving ahead of it by 60 degrees.

There might be many such colonies of varying size. The smaller ones with lengths of a kilometer or so and widths of a tenth of a kilometer might house a population of thousands, while those which are 30 kilometers long and 3 kilometers wide might house millions.

It is difficult to judge whether it is more feasible, from

a political, economic, and psychological standpoint (not technological), to build the space stations directly by Earth labor and using the Moon as a quarry, or to wait for a functioning Lunar colony to take the matter in hand.

My own feeling is that the intermediate production of a Lunar colony would work better. The Moon is a world, and mankind is presently accustomed to worlds. Moon colonies can begin very small indeed, whereas a space colony would probably have to be fairly large, to begin with, in order to supply a sufficient thickness of material to protect the population against cosmic-ray bombardment. (The Lunar crust supplies that protection for even the smallest Lunar colony.)

This means that although a functioning Lunar colony might cost more than a space colony, or even several space colonies, all told, the expense might be spread out over a long time interval and be less painful.

Once the Lunar colony is flourishing, it would be populated by human beings far more space-minded than our Earth population and far readier, psychologically, to build space colonies.

Of course, whether built directly by Earth labor or eventually by Lunar-colonist labor, the end result might be the same. The ships that would explore, exploit, and colonize the Solar system might have either a Lunar-colonist crew or a space-colonist crew.

The point of discussion here, however, is whether the existence of space colonies would make stellar exploration more feasible than the existence of a Lunar colony (and of colonies on other worlds of the Solar system) would.

Consider that a space colony is, in actual fact, the starships we have been discussing; except that the space colonies lack certain of the properties that place difficulties in the way of exploration beyond the Solar system.

First, the initial space colonies, located at the Trojan points of the Earth-Moon system, would remain within the Solar system and, indeed, very close to home and to the oldest and most advanced populated worlds in existence. They would not be moving off into infinite space forever, so that the inhibiting factor of "Goodbye forever" would not exist. The space colonists would not be abashed and distressed by the prospect of lifelong isolation for themselves and their descendants, and the Lunar colonists would not have to watch their entire investment disappear forever. This means that the way of life of a starship could be worked out in comfort and security.

For instance, there would still be the matter of adjusting to the markedly variable gravitational effects, but this could now be done without the added difficulties of breaking the unbilical cord. Those who could not manage to adjust to a variable gravitational intensity could return to the Moon for rest and rehabilitation before trying again— or could give it up forever. It may be that the very knowledge that one could give up if one wished would make it easier not to give up at all.

In fact, we might suppose that variable gravity, once adapted to, would have its own joys. The mountains built along the inner walls of the far ends of a large space colony would offer a new kind of mountain climbing where the higher one climbed, the easier it would be to climb still higher. There might be the joys of personal flying, with wings strapped to one's arms, if one rose to a level near enough to the axis. There could be fun with zero gravity at the axis.

In short, the space colonies would offer a way of developing, in stages and with security, a population thoroughly adapted to the starship way of life.

Then, too, these space colonies would differ from the inhabited natural worlds with respect to mobility. The Earth, the Moon, the various other inhabited bodies of the Solar system, would all be too massive to respond easily to thrusts designed to move them far out of their orbit. It would take an extraordinary and quite prohibitive quantity of energy to move them out of their orbits, largely because of the overwhelming fraction of their bodies that has no connection with life except to serve as inertia-originating mass.

A space colony, which has insignificant mass by world standards, could be moved out of orbit with a comparatively small output of energy. It *could* take off for the stars, if it wanted to, and it would have a new and vast advantage over even the Lunar colonists when it came to making the journey.

A Lunar colonist, in transferring from his home on the Moon to a spaceship, is making a transfer from a larger place to a smaller place of the same sort and that gives him an advantage over the inhabitants of Earth. Nevertheless, he is still leaving home. A space colonist, however, has a further advantage since he makes no transfer at all! His *home* moves. He can travel to the end of the Universe without budging from his own familiar surroundings.

The fact that this is so may still not make it easy for him to leave the Solar system. Although a space colonist has broken away from Earth and Moon alike—from all worlds—he has not broken away from the Sun. If the space colonies are indeed built as part of the Earth-Moon system to begin with, then all will consider the Sun—an object with the appearance in the sky of *our* Sun—to be part of home.

Something else is still needed to make stellar exploration feasible and that something else might come about quite inevitably.

The space colonies would be bound to feel a kind of population pressure. The success of one space colony will make it that much more pressing or urgent—or, in any case, desirable—to build another. With room made available for millions of people on space colonies, the human population (which might have been stabilizing its numbers or even bringing them down) might expand again. Populations, after all, grow to fill the available room and more; this has always happened.

To have a number of space colonies at each Trojan point would be pleasurable. Space colonists could travel from one to another with virtually no expenditure of energy, and each space colony would seem novel and interesting to the visiting tourists from neighbor colonies.

With multiplication, however, there will inevitably come to be too many. Orbital perturbations will limit how many space colonies can safely he crowded into each Trojan point, the Lunar colonists may balk at having to spend more and more of their time, effort, and resources on endless production of space colonies, and the space colonies themselves will feel unpleasantly crowded.

The push will be on for the space colonies to move outward in search of empty space and of new resources for the building of daughter colonies of their own.

The logical way out would be to move to the asteroid belt, where some hundred thousand asteroids can supply the necessary materials (including volatile matter on some of the larger ones, perhaps). The total mass of the asteroids is considerably less than that of the Moon, but that mass is broken into small pieces each with a

248

negligible gravitational effect, so that they would be easier to mine and refashion with techniques that would be far advanced over what they were when the first space colonies came into being.

One difficulty involved in this move is that the Sun is, to an extent, left behind. In the asteroid belt, its apparent size and its actual radiation intensity is brought down to less than a sixth what it is in the Earth-Moon system. This might create an energy problem, but we can be reasonably sure that by the time the move to the asteroids has been made, nuclear fusion will be practical and the Sun can be dispensed with, if necessary, as an energy source—as it will have been on any settled worlds beyond Mars.

The shift from the Earth-Moon system to the asteroid belt will increase the probability that the space colonies will become starships outright, for the following reasons.

1. The mere fact that the Sun is much farther off, much smaller, no longer the source of energy; it will bulk less importantly in the minds and consciousness of the colonists. They will have in this way already taken a step toward freeing themselves of the Solar system generally, as once their ancestors had freed themselves of the Moon, and as earlier ancestors had freed themselves of Earth.

2. Having moved farther from the Sun, less energy will be required for a starship to escape from the Solar system.

Eventually, some space colony, seeing no value in circling round and round the Sun forever, will make use of some advanced propulsion system to break out of orbit and to carry its structure, its contents of soil, water, air, plants, animals, and people, out into the unknown. Except for the fact that the Sun will be shrinking in apparent size, and that radio contact will become steadily more difficult to maintain, until both Sun and radio contact disappear altogether, there will be no difference whatever detectable in the colony as a result of this change in motion.

—But it will have become a starship; and others would undoubtedly follow.

As many space colonies become starships, they will lose contact with each other, but in the vastness of the asteroid zone (as opposed to the constricted neighborhood of the Earth-Moon system) such contact will have been minimal perhaps, in any case.

And in exchange for the loss of the Sun and of contact with other populated objects, there would come the satisfaction of curiosity, of the basic itching desire to know.

Why not see what the Universe looks like? What's out there, anyway? There would also be the satisfaction of the desire for freedom. There might well be an exaltation in being at last an independent component of the Universe, free of stars and planets alike.

Nor need one, perhaps, fear the slow loss of resources through imperfect cycling (what can be perfect?). Once a starship is ready for the ultimate trip, there will be no lack of resources. They are widely spread out, but the starship can travel for several years to pick up a needed item.

They may, for instance, work their way through the comet cloud at the very rim of the Solar system, watching for one of the hundred billion comets present there in its native form as a small body of frozen volatiles. One can be found, picked up, and placed in tow, to serve as a long-time source of hydrogen to keep the fusion reactor running. Given time, and the starship will have nothing in greater profusion than time, a string of them can be picked up.

Nor need the Universe be considered empty once the comet cloud is left behind. It may take hundreds or even thousands of years to reach a star, but it is probably a mistake to assume that the Universe is made up only of stars and their immediate attendants. It happens to be only the stars that we see because they glow brightly, but there are some indications that the Universe as a whole may be considerably more massive than the sum of the masses of the stars. At least part of the additional mass may be small, dark bodies fleeting through the volume of space between the stars.

It may be that no decade will pass without the detection of one or two of these small bodies (and occasionally a fairly large one) that may contain enough involatile material to give the starship an opportunity to flesh out its stores of aluminum, steel, and silicates.

Eventually—a long eventually—many generations, perhaps, the starship may approach a star. It might not be accident. Undoubtedly the ship's astronomers will study all stars within so many light-years' distance, choose one that has a high probability of containing habitable worlds, and head for that.

It might be delightful to land on a world and to experience something remembered only in the mists of legend. If the planet can be burrowed into as the Moon or Mars had once been; or if it has a compatible ecology and no dangerous life forms and can be inhabited on the outside; it may even be tempting to remain there indefinitely, abandoning the starship, which, despite all repairs, might by then be rather battered.

There would be a loss of freedom in doing so, for once again the descendants of the original planet-bound humans would be planet-bound, once again doomed to be perpetually pinned to the planetary surface under the light of a Sun. —Though the novelty might seem so delightful that it would be a long time before the inhabitants would come to bemoan lost freedom.

And there would be a gain, too. Through all the centuries of starship travel, population would have had to be rigidly controlled, resources rigidly conserved and recycled. Now, for a time at least, human beings could spread over the face of a planet, glorying in explosive breeding and expansive wasting. Someday, of course, the time would come when space colonies would be formed again and these converted into starships for the renewal and continuation of the vast exploration.

We might almost imagine human societies as existing in two alternating forms: a motile, population-controlled form in starships drifting through space; and a sessile, population-expanding form on planets circling Suns.

The long separation of various groups of human beings from other groups would encourage cultural and even biological variations that would produce an infinite richness of experience and culture; richness that could not conceivably be duplicated on a single world or in a single planetary system.

Competing cultures might have a chance to interact when the paths of two starships intersected.

Detecting each other from a long distance, the approach of two starships might be a time of great excitement on each. The meeting would involve a ritual of incomparable importance; there would be no flash-by with a hail and farewell.

Each, after all, would have compiled its own records, which it could now make available to the other. There would be descriptions, by each, of sectors of space never visited by the other. New theories and novel interpretations of old ones would be expounded. Literature and works of art could be exchanged, differences in custom explained.

Most of all, there would be the opportunity for a cross-flow of genes. An exchange of population (either temporary or permanent) might be the major accomplishment of any such meeting. And if the separation had been so long-enduring that the two groups had evolved into separate species, mutually infertile, there would be an intellectual cross-fertilization at any rate.

In this way, mankind would become no longer a creature of Earth or the Solar system, but of the whole Universe, drifting outward, ever outward, perhaps even out of our galaxy and into others, until such time as the Universe finally came to an enormously slow end and through one route or another could no longer support life.

But wait, we have no right to consider ourselves the only intelligent beings in the Universe. In our own galaxy there may be hundreds of millions of Earth-like planets, each with its load of life. And there may be similar numbers in each of the many billions of other galaxies. Even if only on one life-bearing planet out of millions, intelligent life forms evolve and form technological civilizations, there may still be millions of such civilizations in the Universe.

Every one of these may develop to the point where (like mankind today) they stand in danger of being killed by their own success. Those who survive that danger (as we may perhaps—but not certainly—survive it) may go on, more or less inexorably, to the starship stage.

Depending on when an intelligent species got its start and how rapidly it developed, the starships may have been launched, in some cases, many millions of years ago, while others will be launched many millions of years from now.

As to those already launched, we might wonder why they have not reached us.

That may be because the vast distances of the Universe insulate us. In the entire lifetime of the galaxies so far, it may be that starships have penetrated and explored only a fantastically small percentage of the whole. Or else they may have missed us through sheer chance; passing through our sector of the Universe, they may simply never have come close enough to us to observe or be observed. Or else they may know of our existence, but choose to leave us alone as part of some basic ethic that all intelligent species have the right to develop undisturbed.

It may be, though, that on some wholly fortunate day, a human starship will encounter an alien starship. A genetic interchange would then be out of the question of course, but the intellectual cross-fertilization would be incredibly richer than in the case of a human-human interchange, assuming the human-alien transfer could be made at all.

Surely, by that point in history, it will be understood that it is the nature of the mind that makes individuals kin, and that the differences in the shape, form, or manner of the material atoms out of whose intricate interrelationships that mind is built are altogether trivial.

It may be that as the human starships start moving outward—with what advances in human knowledge and development we can scarcely imagine—they may eventually find themselves to be part of a vast brotherhood of intelligence; a complex of the innumerable routes by which the Universe has evolved in order to become capable of understanding itself.

And as part of this brotherhood, spanning and filling the Universe, mankind will have found its true goal at last.

AFTERWORD

The rather vast picture of the future with which the preceding essay closes seems a good point at which to leave you all, so farewell.

Nevertheless, barring acts of God, I rather think I'll be back, and I hope you'll be joining me again.